This book deals with the effect of crystal symmetry in determining the tensor properties of crystals. Although this is a well-established subject, the author provides a new approach using group theory and, in particular, the method of symmetry coordinates, which has not been used in any previous book.

Using this approach, all tensors of a given rank and type can be handled together, even when they involve very different physical phenomena. Applications to technologically important phenomena as diverse as the electro-optic, piezoelectric, photoelastic, piezomagnetic, and piezoresistance effects, as well as magnetothermoelectric power and third-order elastic constants are presented. Attention is also given to 'special magnetic properties', that is those that require the concepts of time reversal and magnetic symmetry, an important subject not always covered in other books in this area.

This book will be of interest to researchers in solid-state physics and materials science, and will also be suitable as a text for graduate students in physics and engineering taking courses in solid-state physics.

CRYSTAL PROPERTIES VIA GROUP THEORY

CRYSTAL PROPERTIES VIA GROUP THEORY

ARTHUR S. NOWICK

Henry Marion Howe Professor Emeritus of Materials Science
Columbia University, New York

CAMBRIDGE
UNIVERSITY PRESS

CAMBRIDGE UNIVERSITY PRESS
Cambridge, New York, Melbourne, Madrid, Cape Town, Singapore, São Paulo

Cambridge University Press
The Edinburgh Building, Cambridge CB2 2RU, UK

Published in the United States of America by Cambridge University Press, New York

www.cambridge.org
Information on this title: www.cambridge.org/9780521419451

First published 1995
This digitally printed first paperback version 2005

A catalogue record for this publication is available from the British Library

ISBN-13 978-0-521-41945-1 hardback
ISBN-10 0-521-41945-X hardback

ISBN-13 978-0-521-02231-6 paperback
ISBN-10 0-521-02231-2 paperback

To Joan,
the source of my inspiration
and to symmetry,
a source of beauty in our lives

Contents

Preface

The study of the anisotropic properties of crystals, often called 'Crystal Physics', is the oldest branch of solid-state physics, dating back to the turn of the twentieth century and the treatises of W. Voigt. It deals with the 'matter tensors' that describe such anisotropic properties, and the way that these tensors are simplified as a result of the existence of crystal symmetry. In recent years, there have been many textbooks on this subject. Most widely known is that by J. F. Nye (*Physical Properties of Crystals*, Oxford University Press, 1957), who introduced matrices and tensors to create a more compact notation than that used earlier, but did not use group theory.

Group theory provides the ideal mathematical tools for dealing with these problems elegantly and compactly. These methods have been used by various authors, notably Fumi, Bhagavantum and Juretshke. However, the usefulness of group theory was not always recognized. In fact Nye (page 122 of his book), commenting on work using group theory, states: 'group theory . . . does not reveal which moduli are independent but only the total number of independent ones'. The present book is dedicated to showing, not only that this statement is untrue, but that the use of group theory lends elegance and beauty to what would otherwise be dull calculations. In this book we utilize the method of symmetry coordinates, very much as is used in the study of molecular vibrations (e.g. as described in the book by Wilson, Decius and Cross). This method has not been employed in other books on the present subject, even those that utilize group theory.

The plan of the book is as follows. In the first two chapters we introduce the full range of properties to be studied (electrical, magnetic, mechanical, dielectric, optical, transport, etc.), including the concept of matter tensors. Here we distinguish between equilibrium properties (Chapter 1) and transport properties (Chapter 2), each of which has a different theoretical basis. The tensors introduced are classified by their ranks and intrinsic symmetries. This scheme, whereby the properties are handled separately from the questions of crystal symmetry, is unique to the present book. Next, in Chapter 3, we introduce the concepts of crystal symmetry and of group theory. The reader who has had previous exposure to group theory may treat this chapter as a review, but a reader with no previous background may study it as an introduction to the subject. Indeed he/she will find such a background useful in studying other branches of solid-state physics as well.

In Chapter 4, we show how the method of symmetry coordinates can be applied to the study of matter tensors in crystals. A unique feature is the presentation of the 'Symmetry-Coordinate Transformation tables' for all 32 point groups. These include the requirement of similarity of orientation for those symmetry coordinates that belong to two- or three-dimensional irreducible representations. (To my knowledge, such tables are not given in any other book on group theory.)

The subsequent chapters then deal with the various tensors by rank, rather than by their physical nature. Thus we show that the simplification of matter tensors due to crystal symmetry is a mathematical problem, and may occur in exactly the same way for widely diverse physical properties that have the same tensor rank and intrinsic symmetry. These chapters deal, in ascending complexity, with tensors up to sixth rank (so as to cover the 'third-order elastic constants'). Along the way, 'special magnetic properties', namely, those that require consideration of time reversal and of magnetic point groups in order to analyze them correctly, are handled separately (in Chapters 5 and 8).

Although the subject of the book is a mathematical one, practical examples of materials that show such diverse and technologically important phenomena as the piezoelectric effect, elasticity, the electro-optic effect, photoelasticity, piezoresistance, magnetoresistance and third-order elasticity are discussed. Properties of materials of special interest, such as quartz, the YBCO high-temperature superconductor and lithium niobate are used for illustration, where appropriate.

The material in Chapters 1–8 can comfortably constitute a one-semester course. Problems are given at the end of each chapter, and a list of general references for further reading, which is combined with the specific references mentioned in the text, is given at the end of the book.

I am particularly grateful to the late William R. Heller, with whom I first carried out research in crystal properties and in defect relaxation phenomena using the present method of symmetry coordinates. He introduced me to this method and was a wonderful collaborator. I am also indebted to my colleague Daniel N. Beshers and to several former graduate students, particularly Wing-Kit Lee, Shiun Ling, Diane S. Richter and Tracey Scherban, all of whom gave critical comments which contributed to improving the manuscript. Finally, I wish to thank A. Continenzà for the quartz-crystal model that appears on the front cover.

A. S. Nowick

1

Tensor properties of crystals: equilibrium properties

A crystal is made up of a regular arrangement of atoms in a pattern that repeats itself in all three spatial dimensions. This structural feature has the practical result that most properties of crystals are anisotropic, that is, different values are obtained for the property when measured along the different directions of a crystal. This anisotropy distinguishes crystals from non-crystalline materials (glasses) or from random polycrystalline aggregates, both of which show isotropic properties.

The purpose of this book is to show how to analyze the anisotropic properties of crystals in terms of the (tensor) nature of the properties and of the symmetry of the crystals. In the first two chapters, we focus on the nature of the various properties with which we will be dealing. We will see how the tensor character of a property helps to define its variation with orientation. Questions of crystal symmetry will then be dealt with, starting from Chapter 3.

1–1 Definition of crystal properties

For a crystal, regarded as a thermodynamic system (i.e. in equilibrium with its surroundings), any physical property can be defined by a relation between two measurable quantities. For example, the property of crystal density is simply the ratio of mass to volume, both measurable quantities; similarly, elastic compliance is the ratio of mechanical strain to stress. Often, one of the measurable quantities can be regarded as a generalized 'force' and the other as a response to that force. Examples of such 'forces' are electric field, magnetic field or mechanical stress, while the corresponding responses are, respectively, electric displacement, magnetic induction and strain. A given response, Y_i, is said to be 'conjugate' to a force X_i if the product $X_i \Delta Y_i$ that accompanies a change ΔY_i in the value of Y_i represents an element of work *per unit volume* done on the system. According to the first and second laws of thermodynamics:

$$\mathrm{d}U = X_i \,\mathrm{d}Y_i + T\,\mathrm{d}S \tag{1-1}$$

where U is the internal energy per unit volume, T the Kelvin temperature and S the entropy per unit volume. In the $X_i \,\mathrm{d}Y_i$ term a summation over i is implied, in accordance with the Einstein summation convention that when a given suffix appears twice in the same term, summation with respect to the suffix is to be understood. (This convention will be used throughout this book.) In view of Eq. (1),* it is also convenient to regard the temperature T to be a 'force' and entropy S as its conjugate response, in which case Eq. (1) may simply be written as

$$\mathrm{d}U = X_i \,\mathrm{d}Y_i \tag{1-2}$$

Although the contributions to U come from conjugate pairs, it does not follow that a force X_i only gives rise to its conjugate response Y_i. In fact, in general, X_i must be considered to give rise to various Y_j. To a first-order approximation, it is often assumed that a given response Y_i is, in fact, *linearly* related to all the forces acting on the material. Thus, we have

$$Y_i = K_{ij} X_j \tag{1-3}$$

This equation assumes implicitly that our reference state is that for which Y_i and X_j are both zero and that all values Y_i and X_j remain small. More generally, we may write linear equations for changes of these variables from reference values Y_{i0} and X_{j0} as

$$Y_i - Y_{i0} = K_{ij}(X_j - X_{j0}) \tag{1-4}$$

(The form of (4) is particularly appropriate for the conjugate variables T and S, for which reference values are not conveniently taken as zero.)

Equation (3) or Eq. (4) represent nothing more than a Taylor expansion of Y_i in terms of all the X_j carried out to terms of first order. (Carrying the expansion further leads to higher-order effects which will be dealt with later in this chapter.) An equation such as Eq. (3) is called a 'constitutive relation', and the quantities K_{ij} represent the physical properties of the crystal. Thus, while the magnitude of a force can be chosen arbitrarily, in order to obtain any given response, the quantity K_{ij}, which is a response per unit force, is a property of the crystal, over which the experimenter has no control once the material is selected.

* Throughout this book, reference to equations in the same chapter will be given without the chapter number, for example, Eq. (1) rather than (1–1); where equation numbers other than those from a current chapter are referred to, these will be preceded by the appropriate chapter number.

An important thermodynamic result is that the matrix of coefficients K_{ij} is symmetric, namely,

$$K_{ij} = K_{ji} \tag{1–5}$$

This result may be proved as follows. We define a thermodynamic potential, Φ, by

$$\Phi \equiv U - X_i Y_i \tag{1–6}$$

Then by combining Eqs. (2) and (6) we obtain for the differential of Φ

$$\mathrm{d}\Phi = -Y_i \,\mathrm{d}X_i \tag{1–7}$$

Since, from thermodynamics, the internal energy function is a state function of all the X_i taken as independent variables, the function Φ is also such a state function. Accordingly, it follows that

$$\partial Y_i/\partial X_j = \partial Y_j/\partial X_i = -(\partial^2\Phi/\partial X_i \partial X_j) \tag{1–8}$$

When this result is compared with Eq. (3) or Eq. (4) it is clear that Eq. (5) follows.

The result of Eq. (5) may be called the *intrinsic symmetry* of the physical property matrix K_{ij}, since it follows from the nature of the property itself and does not require for its validity any symmetry on the part of the crystal under consideration. This intrinsic symmetry is to be contrasted to that imposed by crystalline symmetry, which will be the major subject of the later chapters of this book.

1–2 Physical quantities as tensors; tensor properties

Inherently, a crystal is an anisotropic medium. This means that the response of a crystal to an external 'force' depends not only on the magnitude of that force but also on its orientation relative to the crystal axes. In such cases, the measurable quantities (forces and responses) are components of vectors or tensors, and it is then advantageous to group these components together as appropriate tensor quantities. Thus, for example, the electric field **E** is a vector with three components E_1, E_2 and E_3 which lie parallel to three mutually perpendicular axes. The conjugate response, the electric displacement **D**, is also a vector quantity.* In the absence of any other fields, these two quantities are related by

$$\mathbf{D} = \boldsymbol{\kappa}\mathbf{E} \tag{1–9}$$

* One could equally well take the polarization **P** as conjugate to **E**, the relation between **D** and **P** being: $\mathbf{D} = \kappa_0\mathbf{E} + \mathbf{P}$, where κ_0 is the permittivity of vacuum.

where $\boldsymbol{\kappa}$ is a second-rank symmetric tensor, called the dielectric permittivity tensor, that relates the two vector quantities. (It is symmetric by virtue of Eq. (5).) Thus, just as the forces and responses have been combined in this case to form vectors **E** and **D**, the appropriate crystal properties can be grouped together in the form of a tensor quantity $\boldsymbol{\kappa}$. We may then speak of such quantities as 'tensor properties' of crystals, or simply as 'matter tensors'.

As background to this discussion, for the reader who will find it useful, Appendix A contains a brief review of the theory of tensors. It deals with such questions as the rank of a tensor and the difference between polar and axial tensors.

1–3 The basic linear relations

In order to introduce some important linear relations and the corresponding tensor properties, we consider the following four types of 'forces' and their conjugate 'responses':

Type	'Force'	'Response'
Thermal	Change in temperature, ΔT	Change in entropy/vol., ΔS
Electrical	Electric field, **E**	Electric displacement, **D**
Magnetic	Magnetic field, **H**	Magnetic induction, **B**
Mechanical	Stress, $\boldsymbol{\sigma}$	Strain, $\boldsymbol{\varepsilon}$

The thermal quantities ΔT and ΔS are both scalars, the electrical quantities **E** and **D** are vectors (or polar vectors), the magnetic quantities **H** and **B** are axial vectors, and, finally, the mechanical quantities $\boldsymbol{\sigma}$ and $\boldsymbol{\varepsilon}$ are symmetric second-rank tensors (with six independent components). Thus, tensor types of rank 0–2 are represented in the table above.

If we write the linear constitutive equations in component form, we have

$$\left.\begin{aligned} \Delta S &= (C/T)\Delta T + p_i E_i + q_i H_i + \alpha'_{ij}\sigma_{ij} \\ D_i &= p'_i \Delta T + \kappa_{ij} E_j + \lambda_{ij} H_j + d_{ijk}\sigma_{jk} \\ B_i &= q'_i \Delta T + \lambda'_{ij} E_j + \mu_{ij} H_j + Q_{ijk}\sigma_{jk} \\ \varepsilon_{ij} &= \alpha_{ij}\Delta T + d'_{ijk} E_k + Q'_{ijk} H_k + s_{ijkl}\sigma_{kl} \end{aligned}\right\} \qquad (1\text{–}10)$$

This set of equations, which is nothing more than an explicit form of Eq. (3), covers all possible first-order interactions of the four types of physical quantity presented in the table above. If we wish to take advantage of the tensor character of the quantities involved, we may group the 'forces' and 'responses' together as tensors and write the K_{ij} coefficients in the following block form:

C/T	$\mathbf{p}$	$\mathbf{q}$	$\boldsymbol{\alpha}'$
$\mathbf{p}'$	$\boldsymbol{\kappa}$	$\boldsymbol{\lambda}$	$\mathbf{d}$
$\mathbf{q}'$	$\boldsymbol{\lambda}'$	$\boldsymbol{\mu}$	$\mathbf{Q}$
$\boldsymbol{\alpha}$	$\mathbf{d}'$	$\mathbf{Q}'$	$\mathbf{s}$

Here the matrices $\boldsymbol{\kappa}$ and $\boldsymbol{\mu}$ have dimensions 3×3, $\mathbf{s}$ is 6×6, $\mathbf{q}$ is 3×1, and so on. The diagonal line drawn through the blocks is the symmetry line, about which the relation $K_{ij} = K_{ji}$, that is, Eq. (5), is applicable. The diagonal blocks: C/T, $\boldsymbol{\kappa}$, $\boldsymbol{\mu}$ and $\mathbf{s}$, are called *principal effects*, since they relate a given response tensor to its conjugate 'force' tensor. The off-diagonal blocks are called *cross effects* or *interaction effects*. They show that a given 'force' may produce non-conjugate responses. Not all cross effects need to occur for all crystals, in the sense of being significant in magnitude, but they must all be anticipated in order to allow for complete generality.

We next identify all the quantities in these equations, beginning with the principal (diagonal) matter tensors. Thus C is the specific heat per unit volume at constant fields $\mathbf{E}$, $\mathbf{H}$ and $\boldsymbol{\sigma}$ and, as such, is a generalization of the usual specific heat at constant pressure. It is a scalar crystal property since it relates two scalar quantities. The quantities κ_{ij} are components of the dielectric permittivity tensor already given by Eq. (9). This tensor relates two polar vectors and is therefore a tensor of second rank. Further, it is symmetric by virtue of Eq. (5). The quantities μ_{ij} are components of the magnetic permeability tensor $\boldsymbol{\mu}$ which relates $\mathbf{B}$ to $\mathbf{H}$, that is,

$$\mathbf{B} = \boldsymbol{\mu}\mathbf{H} \tag{1–11}$$

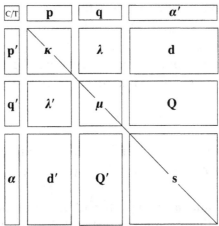

The tensor $\boldsymbol{\mu}$ relates two axial vectors and thus transforms as a (polar) second-rank tensor. (The reason is that any transformation that changes the sign of $\mathbf{H}$ must do the same for $\mathbf{B}$. See Appendix A.) It is also symmetric in view of Eq. (5). Note that these principal effects themselves include off-diagonal components, such as κ_{ij} and μ_{ij} for $i \neq j$.

The last of the principal matter tensors in Eq. (10) are those which relate strain to stress, namely, the elastic compliance constants s_{ijkl}. In Appendix B, we have reviewed the definitions of the six components of strain ε_{ij} and of stress σ_{ij}, both of which are symmetric second-rank tensors, that is, $\varepsilon_{ij} = \varepsilon_{ji}$ and $\sigma_{ij} = \sigma_{ji}$. It is shown there that ε_{11}, ε_{22} and ε_{33} are tensile components while ε_{23}, ε_{31} and ε_{12} are the shears, and similarly for the stress components σ_{ij}. Thus, if we write the generalized Hooke's law:

$$\boldsymbol{\varepsilon} = \mathbf{s}\boldsymbol{\sigma} \tag{1–12}$$

the matrix **s** whose components are s_{ijkl} represents a fourth-rank tensor called the *elastic compliance* tensor. It is symmetric in the interchange of i and j, or k and l, and also that of i, j for k, l. (The first two symmetries are valid because $\boldsymbol{\varepsilon}$ and $\boldsymbol{\sigma}$ are symmetric, and the latter because of Eq. (5).) It follows from these symmetries that there are 21 components in this elastic compliance tensor. If Eq. (12) is inverted, we may write

$$\boldsymbol{\sigma} = \mathbf{c}\boldsymbol{\varepsilon} \tag{1–13}$$

in which **c** is a fourth-rank tensor called the *elastic stiffness* tensor, which has the same intrinsic symmetry and number of independent coefficients as the **s** tensor.

We now turn to the off-diagonal or cross effects, keeping in mind the symmetry derived from thermodynamic considerations, Eq. (5). The coefficients p_i, which represent the entropy (or heat) produced by an electric field, are components of the *electrocaloric effect*, a matter tensor which is a vector quantity since it couples a scalar response to a vector field. The converse effect, p_i', giving the electric displacement caused by a temperature change is called the *pyroelectric effect*. From the reciprocity relations, Eq. (5), it is clear that $p_i' = p_i$. Similarly, the quantities q_i constituting the *magnetocaloric effect* are components of an axial vector, as are the components q_i' of the converse effect, called the *pyromagnetic effect*, with $q_i' = q_i$.

The cross effect α_{ij}, which relates strain to ΔT, constitutes a set of *thermal expansion* coefficients. In general, there are six coefficients, meaning that, in an arbitrary crystal, a change in temperature can produce not only tensile but also shear deformations as well. Since the set of coefficients α_{ij} comprise the components of a strain, they constitute a second-rank symmetric matter tensor. The converse coefficients α_{ij}' constitute the *piezocaloric effect*, and by Eq. (5) and the symmetry of α_{ij} we must have

$$\alpha'_{ij} = \alpha_{ji} = \alpha_{ij} \tag{1–14}$$

so that the direct and converse coefficients are equal.

Still another cross-effect is represented by the coefficients λ_{ij} which relate a polar vector (**D**) to an axial vector (**H**). The quantity $\boldsymbol{\lambda}$, called the *magnetoelectric polarizability* tensor, is then a second-rank *axial* tensor, and it is not symmetric. (This can easily be seen from its position with respect to the diagonal in the block form on page 5.) Thus $\boldsymbol{\lambda}$ has, in general, nine components. The coefficients of the converse tensor λ'_{ij} must obey

$$\lambda'_{ij} = \lambda_{ji} \tag{1–15}$$

as a consequence of Eq. (5).

Next we turn to interactions between electrical and mechanical properties, the so-called 'piezoelectric effects'. The coefficients d_{ijk} relate to the electric displacement produced in a crystal subject to stress; as such they couple a second-rank tensor to a polar vector and must therefore form a tensor of third rank. Since stress is symmetric in its indices, d_{ijk} must be symmetric in the interchange of j and k. The existence of d_{ijk}, the *direct piezoelectric effect* implies (by the reciprocity relations) the existence of a converse effect, that is, that application of an electric field will give rise to an appropriate strain. From Eq. (5) the quantitative relationship between the *converse piezoelectric effect* d'_{pqr} and the direct effect is

$$d'_{pqr} = d_{rpq} \tag{1–16}$$

In addition, both the direct and converse quantities are symmetric in the interchange of p and q in this equation, leading to a total of 18 piezoelectric coefficients in the most general case.

The final cross effect which appears in Eqs. (10) is the *piezomagnetic effect* Q_{ijk} and its converse Q'_{pqr}. These matrices each relate a second-rank symmetric tensor to an axial vector and thus constitute a pair of third-rank axial tensors. As in the case of the piezoelectric effect, Q_{ijk} is symmetric in the interchange of j and k and is related to the converse effect by

$$Q'_{pqr} = Q_{rpq} \tag{1–17}$$

Q'_{pqr} is often called the *magnetostrictive tensor*.

The various equalities of cross effects that follow from Eq. (5), such as those given by Eqs. (14)–(17), have the qualitative meaning that if a force X_j produces a response Y_i, a force X_i will of necessity produce a response Y_j. In addition, however, such equations give a *quantitative* relationship between what might have appeared to be unrelated effects. The importance of these quantitative thermodynamic relationships for the experimenter is that it allows him/her to choose the measurement that is easier to carry out for evaluating a given cross-effect. For example, in piezoelectricity, it may be easier to measure the direct effect (electrical response produced by stress) than the converse effect (strain resulting from an electric field). It is reassuring that, by measuring such a property, the converse property is obtained as well.

Table 1–1 (located at the end of the chapter on page 18) lists these various matter tensors, as well as others which will be introduced later in this chapter. Included in this table (second column) is the tensor character of each property using notation $T(n)$ for a tensor of rank n, a superscript ax for axial quantities and a subscript S to denote that it is symmetric (the fifth column showing exactly what intrinsic symmetry exists among the coefficients). Thus $T(0)$, $T(1)$ and $T(1)^{ax}$ denote, respectively, a scalar, polar vector and axial vector property, respectively. The third column shows the tensor characters of the two measurable quantities that the given matter tensor relates. The final column gives the maximum number of independent components of the property in question, based on its intrinsic symmetry but in the absence of any crystal symmetry. The reader should recall that the intrinsic symmetry can come from either the definition of the property in question or the thermodynamic reciprocity relations of the type of Eq. (5).

1–4 Condensation of indices: the 'engineering' stresses and strains

In much of the solid-state and engineering literature the awkwardness of the need for double indices to specify components of stress and strain is avoided by using a single-index notation running from 1 to 6 instead of from 1 to 3. The convention, which yields the so-called 'engineering' stresses and strains, converts σ_{ij} into σ_k^{e} (the superscript e standing for 'engineering') in the following way:

ij:	11	22	33	23	31	12
k:	1	2	3	4	5	6

in which $k = 1, 2, 3$ represents the tensile components and 4, 5, 6 the shear components. Thus, $\sigma_1^{\text{e}} = \sigma_{11}$, and $\sigma_4^{\text{e}} = \sigma_{23}$, etc. In the case of the strains, however, the conversion of indices is the same, but additional factors of 2 are introduced in the shear strains, as follows:

$$\varepsilon_1^{\text{e}} = \varepsilon_{11}, \qquad \varepsilon_2^{\text{e}} = \varepsilon_{22}, \qquad \varepsilon_3^{\text{e}} = \varepsilon_{33}$$
$$\varepsilon_4^{\text{e}} = 2\varepsilon_{23}, \qquad \varepsilon_5^{\text{e}} = 2\varepsilon_{31}, \qquad \varepsilon_6^{\text{e}} = 2\varepsilon_{12}$$

In this way, two-index compliances are introduced in place of the four-index quantities of Eq. (10):

$$\varepsilon_i^e = s_{ij}^e \sigma_j^e \tag{1–18}$$

and for the stiffness constants:

$$\sigma_i^e = c_{ij}^e \varepsilon_j^e \tag{1–19}$$

the summations now being from 1 to 6. Thus s_{ij}^e and c_{ij}^e are 6×6 matrices which are symmetric as a consequence of Eq. (5). This condensation of indices for stress and strain may also be extended to the piezoelectric and the piezomagnetic coefficients.

The entire set of Eqs. (10) may then be rewritten in terms of these condensed indices, with the introduction of appropriate engineering quantities, and allowing all indices for stresses and strains to run from 1 to 6:

$$\left.\begin{aligned} \Delta S &= (C/T)\Delta T + p_i E_i + q_i H_i + \alpha_i^e \sigma_i^e \\ D_i &= p_i \Delta T + \kappa_{ij} E_j + \lambda_{ij} H_j + d_{ij}^e \sigma_j^e \\ B_i &= q_i \Delta T + \lambda_{ji} E_j + \mu_{ij} H_j + Q_{ij}^e \sigma_j^e \\ \varepsilon_i^e &= \alpha_i^e \Delta T + d_{ji}^e E_j + Q_{ji}^e H_j + s_{ij}^e \sigma_j^e \end{aligned}\right\} \tag{1–20}$$

In this equation, we have also taken advantage of reciprocity relations already discussed that relate the primed physical properties to the unprimed.

For later use, it is advantageous to give the conversions here between the original tensor properties of Eq. (10) and the corresponding 'engineering' properties in Eq. (20), taking cognizance of the factor of 2 which entered in the definition of the engineering shear strains. Thus,

$$\begin{aligned} s_{ijkl} &= s_{mn}^e \quad \text{when } m \text{ and } n \text{ are } 1, 2 \text{ or } 3 \\ 2s_{ijkl} &= s_{mn}^e \quad \text{when either } m \text{ or } n \text{ is } 4, 5 \text{ or } 6 \\ 4s_{ijkl} &= s_{mn}^e \quad \text{when } m \text{ and } n \text{ are } 4, 5 \text{ or } 6 \\ \alpha_{ij} &= \alpha_k^e \quad \text{when } k = 1, 2 \text{ or } 3 \\ 2\alpha_{ij} &= \alpha_k^e \quad \text{when } k = 4, 5 \text{ or } 6 \\ d_{ijk} &= d_{il}^e \quad \text{when } l = 1, 2 \text{ or } 3 \\ 2d_{ijk} &= d_{il}^e \quad \text{when } l = 4, 5 \text{ or } 6 \end{aligned}$$

while $Q_{ijk} \rightarrow Q_{il}^e$ obeys the same relations as for $d_{ijk} \rightarrow d_{il}^e$. Note that the two-index matrices $\mathbf{d}^e$ and $\mathbf{Q}^e$ are 3×6 (non-square) matrices. There are no factors of 2 and 4 in the conversion of the tensor stiffness constants c_{ijkl} to engineering quantities c_{mn}^e.

It is important to note that, while the set of Eqs. (20) looks simpler than that of Eqs. (10), the apparent simplicity is deceiving, because in making the condensation of indices, we have given up our ability to easily handle the transformation of stresses and strains under a change of axes (as given, for example, in Appendix A). In other words, the proper tensor character of the crystal properties has been lost in the engineering notation. We will see later that group theory provides ways of obtaining the advantage of both condensed indices and knowledge of the transformation properties by using appropriate six-vectors for stress and strain. For the present, however, we need to introduce the engineering notation because, in the literature, it is customary to give numerical values of these properties based on this notation rather than on the tensor notation.

1–5 Effect of changing the conditions of measurement

In Eqs. (2) and (10), we have taken the 'forces' as independent variables and the 'responses' as dependent. Thus, the meaning of any quantity whose definition does not involve a given force is that it is to be measured with that force held constant. For example, the dielectric permittivity κ_{ij} in Eq. (10) is to be thought of as measured at constant values of T, **H** and $\boldsymbol{\sigma}$ as well as those of E_k for $k \neq j$. Similarly, the quantity d_{ijk} is $\partial D_i/\partial\sigma_{jk}$ for constant values of T, **E**, **H** and stress components other than jk.

Now, for each conjugate pair of quantities listed in the table on page 4, we may choose either the 'force' or the 'response' as the independent variable, the other being regarded as dependent. Sets of linear constitutive equations similar to Eqs. (10) may then be written in terms of the selected independent variables. In this way we can produce many variations of Eqs. (10), each with coefficients having different meanings. However, any one such set of equations contains, in principle, all the physical information from which the other sets can be derived.

The relationship between coefficients with different quantities held constant can be derived thermodynamically. First we illustrate, with a principal effect, the elastic compliances. Modern measurements of elastic constants often use high-frequency sound-wave propagation (Truell *et al.*, 1969), leaving no time for heat flow into or out of a strained region. Therefore, changes are adiabatic (constant S) rather than isothermal. Writing the thermoelastic parts of Eq. (10) only, we have

$$\left.\begin{aligned} \Delta S &= (C^\sigma/T)\Delta T + \alpha_{ij}\sigma_{ij} \\ \varepsilon_{ij} &= \alpha_{ij}\Delta T + s^{\mathrm{T}}_{ijkl}\sigma_{kl} \end{aligned}\right\} \qquad (1\text{–}21)$$

where C^σ is specific heat at constant stress and superscript T indicates that the constants are isothermal. To obtain adiabatic constants, we set $\Delta S = 0$

in the first of Eqs. (21) and use this result to eliminate ΔT in the second equation to obtain

$$(\partial \varepsilon_{ij}/\partial \sigma_{kl})|_S = s^{S}_{ijkl} = s^{T}_{ijkl} - \alpha_{ij}\alpha_{kl}(T/C^{\sigma}) \qquad (1\text{–}22)$$

Thus the adiabatic elastic compliance s^{S}_{ijkl} equals the isothermal compliance minus a correction term which involves the thermal expansion coefficients (thermoelastic coupling coefficients). This correction term is only ~1% of s^{T}_{ijkl}, as is commonly the case for such corrections for a change in independent variable. More importantly, we note that the tensor character of s^{S}_{ijkl} and of s^{T}_{ijkl} must be the same, as verified by the fact that the two are related by an additive relation in which the correction term is also a fourth-rank tensor with the same intrinsic symmetries as the quantity s^{T}_{ijkl}.

An example of a cross effect is d^{S}_{ijk}, which is easily shown to be given by

$$d^{S}_{ijk} = d^{T}_{ijk} - p_i\alpha_{jk}(T/C^{\sigma}) \qquad (1\text{–}23)$$

in which the correction term now involves both the thermoelectric and thermoelastic coupling coefficients p_i and α_{jk} respectively.

A change of independent variable can also change the nature (i.e. the dimensionality) of the coefficients. This occurs when there is a change in the measurable quantities that are related by the crystal property. For example, in the case of elastic properties, if the strains are taken as independent variables, the crystal-property coefficients that are relevant become the stiffnesses c_{ijkl} (as in Eq. (13)) which are reciprocal to the compliance constants (see Appendix B). For the piezoelectric coefficients, we now have new quantities:

$$e^{T}_{ijk} = (\partial D_i/\partial \varepsilon_{jk})|_{E,T} = -(\partial \sigma_{jk}/\partial E_i)|_{\varepsilon,T} \qquad (1\text{–}24)$$

The reciprocity relation given in Eq. (24) is derived using the function $\Phi + \varepsilon_{ij}\sigma_{ij}$ as a thermodynamic potential, where Φ is defined by Eq. (6). It can readily be shown that the piezoelectric coefficients e_{ijk} are related to the d_{ijk} through the elastic constant tensor:

$$e_{ijk} = d_{ilm}c_{lmjk} \quad \text{or} \quad d_{ijk} = e_{ilm}s_{lmjk} \qquad (1\text{–}25)$$

The e's, like the d's, are of third rank and are symmetric in indices j and k.

From the viewpoint of this book, the important result is that a given effect (principal or cross effect) has the same tensor character and the same intrinsic symmetry, regardless of which variables are the independent ones. Thus, the effect of crystal symmetry on such a tensor, which we will later derive, will be the same for one set of independent variables as for another.

1–6 Higher-order effects

Equations (10) included only first-order (linear) relations among the four types of physical quantity. We might think of these equations, therefore, as containing the linear terms of a Taylor series expansion. The next step of generalization would then be to include second-order effects. Alternatively, we can say that any of the physical properties defined by Eqs. (10), previously regarded as constant, may in fact be a function of a measurable physical quantity (a force or response), thus generating a new physical property of higher order. Still another viewpoint is that the first-order effects are the various second derivatives of the thermodynamic potential Φ (see Eqs. (7) and (8)), so that second-order effects will be represented by third derivatives of Φ (with respect to the independent variables, say, the 'forces'). The possibilities for second-order effects are so numerous that we prefer to select some of the more important properties individually rather than attempt to write down a complete generalization of Eqs. (10).

1–6–1 Permittivity and optical properties

Let us consider, for example, the dependence of dielectric permittivity κ_{ij} on stress:

$$\partial\kappa_{ij}/\partial\sigma_{kl} = -\partial^3\Phi/\partial E_i\partial E_j\partial\sigma_{kl} = \partial d_{jkl}/\partial E_i \qquad (1\text{–}26)$$

The last equality follows from the fact that $D_i = -\partial\Phi/\partial E_i$ (from Eq. (7)), together with the second of Eqs. (10). Thus the dependence of the dielectric permittivity on stress is related to the dependence of a piezoelectric constant on electric field. We may introduce the appropriate new higher-order matter tensor through the dependence of strain on electric field (i.e. expanding the last of Eqs. (10) to higher order in E_k):

$$\varepsilon_{ij} = d'_{ijk}E_k + \gamma_{ijkl}E_kE_l \qquad (1\text{–}27)$$

Here γ_{ijkl} is an appropriate fourth-rank tensor which is symmetric in i and j and in k and l (but not in the interchange of i, j with k, l). It is called the *electrostrictive tensor*. A magnetostrictive tensor may be defined in a similar way (Bhagavantum, 1966, Section 15–6).

Often, the higher-order effects are too small to be detected by means of dc or low-frequency ac electrical measurements, but are quite detectable through the very high precision of optical measurements. In order to describe such measurements we digress briefly to discuss the indicatrix ellipsoid.

In the region of optical frequencies, the index of refraction, n, is given by (Landau and Lifshitz, 1960): $n = (\kappa/\kappa_0)^{1/2}$, where κ_0 is the dielectric permittivity of the vacuum and κ is that of the medium at optical frequencies (i.e. κ/κ_0 is the dimensionless dielectric constant). Thus, the refractive index must also be represented by a second-rank symmetric tensor with, in general, six independent coefficients (although in itself it is not a tensor). It is convenient to define a second-rank tensor which is the reciprocal of the relative permeability tensor and is called the 'impermeability tensor', $\boldsymbol{\beta}$, by the relation

$$\boldsymbol{\beta} = \kappa_0 \mathbf{E}/\mathbf{D} = \kappa_0/\boldsymbol{\kappa}$$

Further, the components of this tensor, β_{ij}, may be used to define an ellipsoid, called the optical index ellipsoid or *indicatrix* (in the manner described in Appendix A, Section A–5) through the equation

$$\beta_{ij} x_i x_j = 1 \tag{1–28}$$

where the x_i are the three spatial coordinates.

By a transformation of axes, the ellipsoid of Eq. (28) can be brought to the form

$$\beta_1 x_2'^{1} + \beta_2 x_2'^{2} + \beta_3 x_2'^{3} = 1 \tag{1–29}$$

in which x_1', x_2' and x_3' are the *principal axes* and

$$\beta_i = \kappa_0/\kappa_i = 1/n_i^2 \tag{1–30}$$

where the κ_i/κ_0 are the principal dielectric constants and the n_i are the principal refractive indices.

The indicatrix in the general form of Eq. (28) has the following property (Nye, 1957, Chapter XIII and Appendix H). For any direction *OP* regarded as the direction of the optical wave normal, there is a perpendicular central section of the indicatrix which is an ellipse. (See Fig. 1–1.) Then the semiaxes of this ellipse (*OA* and *OB* in Fig. 1–1) give the refractive indices of the two wave fronts normal to *OP* that may be propagated through the crystal. Both of these wave fronts are differently plane polarized. The first, polarized parallel to *OA*, has index of refraction equal to *OA*, and similarly for the second wave, which is polarized parallel to *OB*. Since the indices of refraction, and therefore the velocities, of these two waves are different, the crystal is said to be *birefringent* (doubly refracting). Birefringence is then absent only in the special case of a direction OP whose normal ellipse degenerates into a circle. Further discussion of the indicatrix and of birefringent crystals will be given in Section 6–4–5.

Thus far, our discussion of optical properties has been concerned with a first-order effect. On the other hand, if κ is allowed to depend on stress or

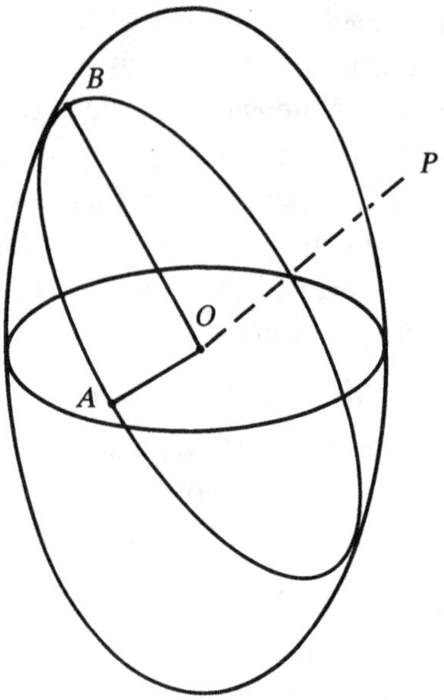

Fig. 1–1. The indicatrix ellipsoid, from which the refractive indices of the two wave fronts normal to direction OP may be obtained.

on electric field, so will the principal refractive indices. Thus, for example, we may write for the changes in coefficients of the impermeability tensor β_{ij} due to such fields

$$\Delta\beta_{ij} = r_{ijk}E_k + p_{ijkl}E_kE_l + q_{ijkl}\sigma_{kl} \tag{1–31}$$

Here r_{ijk} is a third-rank tensor which gives the linear *electro-optic effect* (also known as the Pockels effect), with $r_{ijk} = r_{jik}$, while p_{ijkl} is a fourth-rank tensor representing the quadratic electro-optic or *Kerr effect*. On the other hand, q_{ijkl} is the fourth-rank *photoelastic tensor*. Clearly, both fourth-rank tensors are symmetric in the interchange of i and j and also of k and l. (These properties are also included in Table 1–1.)

These higher-order effects are related to the change in dielectric permittivity with electric and stress fields. Such changes are small and difficult to detect by ordinary low-frequency dielectric measurements, but at optical frequencies they correspond to small changes in the index of refraction which can be measured to great precision. In fact, a difference in refractive index of one part in 10^6 for two waves that are plane polarized at right angles to each other is readily measurable. Nevertheless, the quadratic electro-optic (Kerr) effect is only conveniently studied in crystals for which the linear effect is absent.

The photoelastic effect may, alternatively, be expressed in terms of the strains rather than the stresses, as

$$\Delta\beta_{ij} = m_{ijkl}\varepsilon_{kl} \qquad (1\text{–}32)$$

where the m_{ijkl} are called elasto-optic coefficients and are related to the q_{ijkl} through the elastic constants:

$$m_{ijkl} = q_{ijrs}c_{rskl} \quad \text{or} \quad q_{ijkl} = m_{ijrs}s_{rskl} \qquad (1\text{–}33)$$

In a manner similar to the case of components of stress (see Section 1–4), the six coefficients β_{ij} may be reduced to a one-index 'engineering' notation with

$$\beta_{ij} \rightarrow \beta^{e}_{p} \quad p = 1, 2, \ldots, 6$$

In that case Eq. (31) becomes (omitting the Kerr effect)

$$\Delta\beta^{e}_{p} = r^{e}_{pj}E_j + q^{e}_{pq}\sigma^{e}_{q} \qquad (1\text{–}34)$$

where subscripts p and q run from 1 to 6 and j from 1 to 3. The two-index photoelastic coefficients are then related to the four-index quantities as follows:

$$\begin{aligned} q^{e}_{pq} &= q_{ijkl} && \text{for } q = 1, 2, 3 \\ &= 2q_{ijkl} && q = 4, 5, 6 \end{aligned} \qquad (1\text{–}35)$$

the factor of 2 arising because of pairing of terms σ_{kl} and σ_{lk} when $l \neq k$. Unlike the two-index elastic coefficients, the q^{e}_{pq} are not symmetric with respect to an interchange of p and q, so that, in general, there are 36 independent coefficients.

In a similar way, the elasto-optic coefficients m_{ijkl} may be reduced to two-index notation, except that, now, the factor of 2 does not appear as in Eq. (35). (This is due to the definition of the one-index strains.) Thus Eqs. (33) become

$$m^{e}_{pq} = q^{e}_{pr}c^{e}_{rq} \quad \text{or} \quad q^{e}_{pq} = m^{e}_{pr}s^{e}_{rq} \qquad (1\text{–}36)$$

1–6–2 Third-order elastic constants

Another important higher-order effect comes from the dependence of elastic constants on stress or strain. When dealing with such higher-order effects, 'large deformations' are involved so that we need to use the Lagrangian strains η_{ij} (which reduce to the infinitesimal strains ε_{ij} in the limit of linear elasticity; see Appendix C). One may express the elastic strain energy per unit volume as a power series in the Lagrangian strains:

$$U = U_O + a_{ij}\eta_{ij} + (1/2)c_{ijkl}\eta_{ij}\eta_{kl} + (1/6)C_{ijklmn}\eta_{ij}\eta_{kl}\eta_{mn} + \ldots \qquad (1\text{–}37)$$

The first two terms on the right may be set equal to zero, U_O because it is an arbitrary reference energy and a_{ij} because we carry out the expansion about a state of zero stress. The c_{ijkl} are the elastic constants already discussed. They are called 'second-order' elastic stiffness constants because they are second derivatives of U with respect to the η_{ij}'s; however, they represent a first-order effect in the stress–strain relation. In an analogous way, the coefficients C_{ijklmn} giving rise to the second-order stress–strain effect are called the 'third-order elastic constants' since third derivatives of U are involved in their definition. In a manner analogous to Eq. (31), they may be regarded as the dependence of the second-order elastic constants c_{ijkl} on the strains. The set of these coefficients constitutes a tensor of sixth rank, and these coefficients are symmetric with respect to the interchange of i and j, of k and l, of m and n and/or any of the three pairs ij, kl and mn. In all, there are 56 independent constants. The stresses conjugate to the Lagrangian strains η_{ij}, called the 'thermodynamic tensions', are written as t_{ij} and, in the manner of Eq. (2), are obtained from the energy function U as

$$t_{ij} = \partial U/\partial \eta_{ij} = c_{ijkl}\eta_{kl} + (1/2)C_{ijklmn}\eta_{kl}\eta_{mn} \tag{1–38}$$

As before, we may reduce the indices by defining one-index strains, $\eta_i^{\mathrm{e}}(i = 1, 2, \ldots, 6)$ and writing

$$U = (1/2)c_{ij}^{\mathrm{e}}\eta_i^{\mathrm{e}}\eta_j^{\mathrm{e}} + (1/6)C_{ijk}^{\mathrm{e}}\eta_i^{\mathrm{e}}\eta_j^{\mathrm{e}}\eta_k^{\mathrm{e}} \tag{1–39}$$

The coefficients C_{ijk}^{e} are totally symmetric with respect to all interchanges of i, j and k, but in this form they do not constitute a tensor. Again, there are 56 independent C_{ijk}^{e} for suffixes $i \leqslant j \leqslant k$.

We could also define third-order elastic compliance constants, S_{ijk}^{e}, but it is usually advantageous to deal with the C's, that is, with strains rather than stresses as independent variables. This is because third-order constants are almost always measured by high-frequency pulse techniques, where the strains that are not directly involved in the measurement are the quantities held constant (rather than their conjugate stresses) (Truell *et al.*, 1969).

In a similar way, by expanding Eqs. (37) or (39) out to still higher terms, one can define fourth-order elastic constants. However, the number of coefficients involved in such terms is extremely large and the importance of such quantities is very limited.

1–7 Optical activity: the gyration tensor

In the preceding section, those optical properties were discussed which enabled one to study higher-order effects involving the dielectric constant.

There remains an additional optical property that we wish to discuss, which is not a higher-order effect, namely, the property of optical activity.

When plane polarized light passes through a crystal in a given direction, we may observe rotation of the plane of polarization by an amount proportional to the distance traversed in the crystal, a phenomenon called 'optical activity'. By convention, the sign of the rotation is taken as positive if the plane of polarization is rotated in a clockwise direction as seen by an observer looking into the beam of light. It is easy to see that the sign of the rotation reverses upon reflection through any plane containing the beam. The angle of rotation per unit path length is proportional to a scalar quantity G called the 'gyration'. Now, G is not a true scalar because of the above-mentioned reversal upon reflection; rather it is a pseudoscalar quantity (see Appendix A). Further, the variation of G with direction in the crystal may be written as a quadratic function of the direction cosines, ℓ_i, of the wave normal (i.e. the beam direction) with respect to arbitrarily chosen axes:

$$G = g_{ij}\ell_i\ell_j \tag{1–40}$$

This equation defines a matter tensor g_{ij}, called the *gyration tensor* which, clearly, is symmetric:

$$g_{ij} = g_{ji} \tag{1–41}$$

Equation (40) may be written in vector–tensor notation as

$$G = \ell^{\dagger}\mathbf{g}\ell \quad \text{or } \ell G = \mathbf{g}\ell \tag{1–42}$$

where ℓ, is a polar vector ($\ell^{\dagger}$ its transpose). Since G is a pseudoscalar, ℓG (or $\mathbf{g}\ell$) is an axial vector. Thus, the gyration tensor $\mathbf{g}$ relates an axial vector to a polar vector. It is, therefore, a second-rank symmetric axial tensor. The same conclusion could also have been obtained from Eq. (40) when one considers that ℓ is a vector and G a pseudoscalar.

1–8 Summary of equilibrium properties

We have dealt with a wide range of physical properties: thermal, electrical, magnetic, elastic, and optical, as well as a large number of cross effects among these various types. Specific materials that show these properties will be dealt with in later chapters. The physical phenomena involved in most of these properties have very little relationship to each other. Yet, mathematically, there are common features which will be important in our considerations of the role of crystal symmetry. Most importantly, the tensor character of each physical property is determined

Table 1–1. *Representative equilibrium crystal properties*

Property	Tensor character	Relation between	Symbol	Symmetric with respect to	Maximum no. of constants
Specific heat	$T(0)$	$T(0)$ and $T(0)$	C	—	1
Electrocaloric	$T(1)$	$T(0)$ and $T(1)$	p_i	—	3
*Magnetocaloric	$T(1)^{ax}$	$T(0)$ and $T(1)^{ax}$	q_i	—	3
Thermal expansion		$T(0)$ and $T_S(2)$	α_{ij}		
Dielectric permittivity	$T_S(2)$	or	κ_{ij}	$i \leftrightarrow j$	6
Indicatrix		$T(1)$ and $T(1)$	β_{ij}		
Magnetic permeability	$T_S(2)$	$T(1)^{ax}$ and $T(1)^{ax}$	μ_{ij}	$i \leftrightarrow j$	6
Optical activity	$T_S(2)^{ax}$	$T(1)$ and $T(1)^{ax}$	g_{ij}	$i \leftrightarrow j$	6
*Magnetoelectric polarization	$T(2)^{ax}$	$T(1)$ and $T(1)^{ax}$	λ_{ij}	—	9
Piezoelectricity	$T(3)$	$T(1)$ and $T_S(2)$	d_{ijk}	$j \leftrightarrow k$	18
Electro-optic	$T(3)$	$T_S(2)$ and $T(1)$	r_{ijk}	$i \leftrightarrow j$	18
*Piezomagnetism	$T(3)^{ax}$	$T(1)^{ax}$ and $T_S(2)$	Q_{ijk}	$i \leftrightarrow j$	18
Elasticity	$T_S(4)$	$T_S(2)$ and $T_S(2)$	s_{ijkl}, c_{ijkl}	$i \leftrightarrow j$, $k \leftrightarrow l$, $ij \leftrightarrow kl$	21
Electrostriction	$T(4)$	$T_S(2)$ and $T_S(2)$	γ_{ijkl}	$i \leftrightarrow j$, $k \leftrightarrow l$	36
Photoelasticity			q_{ijkl}		
Kerr effect			p_{ijkl}		
Third-order elasticity	$T_S(6)$	T_S and $[T_S(2)]^2$	C_{ijklmn}	$i \leftrightarrow j$, $k \leftrightarrow l$, $m \leftrightarrow n$, $ij \leftrightarrow kl \leftrightarrow mn$	56

*Special magnetic properties.

by the tensor characters of the two physical quantities that it relates. If the tensor character of two different physical properties is the same, we will see that these properties have much in common mathematically, even though they may be totally unrelated physically.

The major equilibrium properties discussed in this chapter and their principal features and intrinsic symmetries are summarized in Table 1–1 (page 18). Three properties in this table have been singled out with an asterisk as 'special magnetic properties'. These are properties in which components of the magnetic induction, B_i, or of the magnetic field, H_i, appear in one or other of the interacting physical quantities an *odd number of times*. This strange definition will be justified later (in Chapter 5), where it will be shown that to handle the effect of symmetry on such properties requires the introduction of the magnetic crystal classes, which include time reversal as a symmetry operation. Note that, in terms of the above definition, the magnetic permeability, Eq. (11), in which magnetic quantities appear once on either side (therefore, a total of twice) is not a 'special magnetic property'.

Problems

1–1. Obtain Eq. (23) for the relation between the adiabatic and isothermal piezoelectric constants.

1–2. Obtain the reciprocity relation of Eq. (24) and then derive the relations between the two types of piezoelectric constant e_{ijk} and d_{ijk} as given by Eq. (25).

1–3. Derive the relation between the elastic-optic coefficients m_{ijkl} and the photoelastic coefficients q_{ijkl} given by Eq. (33).

1–4. Verify the relation between the two-index and four-index photoelastic coefficients as given by Eq. (35).

1–5. Show that the third-order elastic constants C_{ijklmn} may be written as a dependence of the ordinary (second-order) elastic stiffness constants c_{ijkl} on strain.

1–6. Consider the collection of higher-order effects based on $\partial^3\Phi/\partial E_i \partial H_j \partial \sigma_{kl}$. Show which of these effects are interrelated.

2

Tensor properties of crystals: transport properties

2–1 General theory

Not all physical properties involve systems in thermodynamic equilibrium, in which a measurement of the property can be carried out under reversible conditions. Rather, the properties that we will consider in this chapter involve flow phenomena, that is, systems which undergo small departures from equilibrium and are observed under *steady-state conditions*. Here the 'force' giving rise to the flow is generally the gradient of a scalar quantity, while the 'response' is a flux. The principal examples involve the flow of heat, electricity and mass.

The flux of heat is given by a generalization of Fourier's law:

$$h_i = -K_{ij}\nabla_j T(i, j = 1, 2, 3) \tag{2–1}$$

Here h_i is a component of the flux of energy, that is, heat flowing across a unit area perpendicular to x_i per second, $\nabla_j T = \partial T/\partial x_j$ is the jth component of the temperature gradient, while the K_{ij} are components of the *thermal conductivity tensor*, which represent a physical property of the crystal (a matter tensor). (The minus sign in Eq. (1) indicates that the direction of flow tends to be opposite to that of the thermal gradient.)

The flux of electricity is given by a generalization of Ohm's law:

$$j_i = -\sigma_{ij}\nabla_j\phi = \sigma_{ij}E_j \tag{2–2}$$

where j_i is the component of flux of electrical charge, $\nabla_j\phi$ a component of the gradient of electric potential, and E_j a component of the electric field. The crystal property is now given by the *electrical conductivity tensor* σ_{ij}. Equation (2) may also be inverted to give

$$E_i = \rho_{ik}j_k \tag{2–3}$$

where ρ_{ik} are the components of the *resistivity tensor*.

Finally, mass flow can be expressed by a generalization of Fick's law, the driving 'force' being a concentration gradient $\nabla_j c$ that gives rise to a flux of matter m_i, according to

$$m_i = -D_{ij}\nabla_j c \tag{2–4}$$

Here the crystal property is given by the *diffusivity tensor*, D_{ij}.

That each of these three flow equations involves a tensor crystal property, rather than a scalar quantity, is analogous to the equilibrium properties of Chapter 1, for example Eq. (1–9). It means that the direction of flow in a crystal may be very different from the direction of the gradient that drives the flow. For example, consider, as a special case, a layer-like crystal structure where flow in the layers takes place much more easily than flow perpendicular to the layers. In this case, if a gradient were applied at a large angle (but not 90°) to the layers, the flow direction would still lie very close to the layer planes. Accordingly, the direction of the flux and that of the gradient would be far from antiparallel to each other, as would be the case if they were coupled by a scalar quantity.

Each of the above matter tensors relates a vector quantity (gradient of a scalar field) to another vector quantity (a flux), so that each is a second-rank tensor. Equations (1)–(4), however, all represent what we termed earlier 'principal effects', in this case, the effect of a gradient of a given type (thermal, electrical or matter) in producing flow of the same type. The linear relationship is presumed to apply because the departure from equilibrium is not too great. Nevertheless, in an exactly analogous way to Eqs. (1–3) or (1–10), the most general linear relations must be written allowing for cross (interaction) effects, that is, for the possibility that one type of gradient can also give rise to other types of flux. Such relations are the basis of a general theory, called 'thermodynamics of irreversible processes' about which several treatises have been written (see the references). By analogy to Eq. (1–3) we write, for irreversible processes, the following set of linear equations involving generalized forces X_i and fluxes J_i:

$$J_i = L_{ij}X_j \tag{2–5}$$

Because the system is not in thermodynamic equilibrium, we can no longer use the type of argument given by Eqs. (1–6) to (1–8), which lead to symmetry of the coefficients, Eq. (1–5). However, in the case of transport phenomena there is a different set of arguments leading to the basic theorem of irreversible thermodynamics, called *Onsager's principle*, which states that: provided that the fluxes J_i and forces X_j are chosen correctly, the L_{ij} coefficients are symmetric, or

$$L_{ij} = L_{ji} \tag{2–6}$$

The theoretical basis for Onsager's principle is the 'principle of microscopic reversibility', which states that if the velocities of all particles of a system are simultaneously reversed, they will all retrace their former paths. Details are given in any of the treatises on irreversible thermodynamics. Suffice it to say that Eq. (6) provides the basis for requiring that each of the tensors $\mathbf{K}$, $\boldsymbol{\sigma}$ and $\mathbf{D}$ of the three principal effects be symmetric, for example $K_{ij} = K_{ji}$, etc. Thus, they are all second-rank symmetric tensors, of type $T_S(2)$.

We must now turn to the question of what constitutes correct choice of the forces and fluxes in Eqs. (5). The answer is that the choice must be such that

$$X_i J_i = T\mathrm{d}S/\mathrm{d}t \tag{2–7}$$

where $\mathrm{d}S/\mathrm{d}t$ is the rate at which entropy is produced in the system. (This requirement still leaves some freedom in the proper choice of fluxes and forces.)

2–2 Thermoelectric effects

For our present purposes, the thermoelectric cross effects will be considered. We concern ourselves only with simultaneous heat and electrical flow involving the fluxes h_i and j_i that appeared in Eqs. (1) and (2). It can then be shown, in accordance with the requirements of Eq. (7), that the correct choice for the thermal force is

$$\mathbf{X}^{\mathrm{q}} = -(1/T)\boldsymbol{\nabla} T \tag{2–8}$$

and for the electrical force

$$X^{\mathrm{e}} = -\boldsymbol{\nabla}\bar{\phi} \tag{2–9}$$

where

$$\bar{\phi} \equiv \phi - \mu/e$$

is a generalization of the electrical potential, called the *electrochemical potential*, which includes the chemical potential μ of the electrons. (Here e is the magnitude of the electronic charge, and μ is a function of both composition and temperature.) The specific flow equations analogous to Eqs. (5) may then be written

$$\left.\begin{aligned} j_i &= -\sigma_{ij}\nabla_j\bar{\phi} + \beta_{ij}(\nabla_j T)/T \\ h_i &= \beta'_{ij}\nabla_j\bar{\phi} - \gamma_{ij}(\nabla_j T)/T \end{aligned}\right\} \tag{2–10}$$

Note that, due to the temperature dependence of μ, $\nabla\bar{\phi}$ includes a part that depends on ∇T. In the absence of a temperature gradient, however, $\nabla\bar{\phi}$ becomes just $\nabla\phi$. The quantity σ_{ij} is, therefore, the electrical conductivity tensor of Eq. (2). The Onsager relations, Eq. (6), applied to Eqs. (10) show that

$$\sigma_{ij} = \sigma_{ji}, \quad \gamma_{ij} = \gamma_{ji} \tag{2–11}$$

and also for the cross terms

$$\beta'_{ij} = \beta_{ji} \tag{2–12}$$

The tensors $\boldsymbol{\beta}$ and $\boldsymbol{\beta}'$ are not, in general, symmetric. (See the analogous argument presented in Section 1–3.)

Rather than working with Eqs. (10), it is more useful to write equations in which ∇T and $\mathbf{j}$ are the independent variables. A straightforward conversion of Eqs. (10) (see Problem 2–1), gives (in vector–tensor notation)

$$\left.\begin{aligned} -\nabla\bar{\phi} &= \boldsymbol{\rho}\mathbf{j} - \boldsymbol{\Sigma}\nabla T \\ \mathbf{h} &= -T\boldsymbol{\Sigma}'\mathbf{j} - \mathbf{K}\nabla T \end{aligned}\right\} \tag{2–13}$$

where the cross effect tensor $\boldsymbol{\Sigma}$ is related to $\boldsymbol{\beta}$ by

$$\boldsymbol{\Sigma} = \boldsymbol{\rho}\boldsymbol{\beta}/T \tag{2–14}$$

while

$$\left.\begin{aligned} \boldsymbol{\Sigma}' &= \boldsymbol{\Sigma}^{\dagger} \\ \mathbf{K} &= (1/T)(\boldsymbol{\gamma} - \boldsymbol{\beta}^{\dagger}\boldsymbol{\rho}\boldsymbol{\beta}) \end{aligned}\right\} \tag{2–15}$$

Here $\boldsymbol{\rho}$ is the electrical resistivity tensor (the reciprocal of $\boldsymbol{\sigma}$). The tensor $\mathbf{K}$ in Eq. (13) is the usual thermal conductivity which relates $\mathbf{h}$ to ∇T in the absence of a current flow (rather than in the absence of an electric field).

The off-diagonal tensor, $\boldsymbol{\Sigma}$, which appears in Eqs. (13) is called the *thermoelectric power tensor*. It is a second-rank tensor (relating two vectors) but it is not in general, symmetric. In the first of Eqs. (13), $\boldsymbol{\Sigma}$ gives the effect of a temperature gradient in producing an electrical potential difference in an open circuit (i.e. for $\mathbf{j} = 0$). The difference between two such potential differences across the two members of a thermocouple constitutes the *Seebeck effect*. In the second equation, $\boldsymbol{\Sigma}'$ gives the effect of a current in producing heat evolution or absorption when the temperature is maintained uniform. This is the well-known *Peltier effect*. The first of Eqs. (15), which is a direct consequence of the Onsager relations, then provides an interrelation between these two diverse thermoelectric phenomena.

Note that we could have introduced a third equation, that for mass flow, into Eqs. (10) or (13). This would have introduced additional cross effects, such as the Soret effect, by which a temperature gradient can produce mass flow. However, to avoid excessive complexity we limit ourselves to a single interaction effect for flow phenomena.

Table 2–1, at the end of this chapter, p. 29, lists the various transport properties discussed here, following the same scheme as that used in Table 1–1.

2–3 Piezoresistance

Here we are concerned with a higher-order effect, namely, the effect of stress on the resistivity tensor. In general, we may write

$$\delta\rho_{ij} = \Pi_{ijkl}\sigma_{kl} \tag{2–16}$$

for the change in component ρ_{ij} of the resistivity tensor. The *piezoresistivity tensor* $\mathbf{\Pi}$ is a fourth-rank tensor since it relates two second-rank tensors. It is symmetric in i and j and in k and l (since $\boldsymbol{\rho}$ and $\boldsymbol{\sigma}$ are, respectively, symmetric in these indices) but not in the interchange of i, j for k, l. Because of this symmetry, it is convenient to introduce a one-index notation for $\boldsymbol{\rho}$ and $\boldsymbol{\sigma}$ (following the definition given for the stresses in Section 1–4) and a two-index notation for $\mathbf{\Pi}$:

$$\delta\rho_i^{\mathrm{e}} = \Pi_{ij}^{\mathrm{e}}\sigma_j^{\mathrm{e}}(i, j = 1, 2, \ldots, 6) \tag{2–17}$$

where the superscript e is again used as a reminder that the quantities involved are not proper vector or tensor quantities. Since Π_{ij}^{e} is not symmetric, it contains 36 independent constants, in general.

2–4 Galvanomagnetic and thermomagnetic effects

In the presence of a magnetic field, all the transport phenomena discussed in Section 2–1 occur, except that the physical property tensor components may now be a function of magnetic field, $\mathbf{H}$, or, more conveniently, of magnetic induction, $\mathbf{B}$. In order to deal with such questions we need to apply the generalization of the Onsager relations, Eq. (6), in the presence of magnetic effects. Such a generalization is required because the principle of microscopic reversibility, on which Eq. (6) is based, fails for charged particles moving in a magnetic field, unless the direction of the field as well as the velocities of the particles is reversed. These considerations lead to the modified reciprocity relations:

$$L_{ij}(\mathbf{B}) = L_{ji}(-\mathbf{B}) \tag{2–18}$$

(or similarly, with **H** substituted for **B**). Equation (18) may be applied to each of the transport tensors discussed earlier. For example, in the case of the resistivity,

$$\rho_{ij}(\mathbf{B}) = \rho_{ji}(-\mathbf{B}) \tag{2–19}$$

so that $\boldsymbol{\rho}$ is no longer symmetric. Any second-rank tensor can, however, always be written as the sum of a symmetric and an antisymmetric part:

$$\rho_{ij} = \rho_{ij}^{\mathrm{S}} + \rho_{ij}^{\mathrm{aS}} \tag{2–20}$$

where the superscript denotes the symmetry. By definition,

$$\rho_{ij}^{\mathrm{S}}(\mathbf{B}) = \rho_{ji}^{\mathrm{S}}(\mathbf{B}) \tag{2–21}$$

and

$$\rho_{ij}^{\mathrm{aS}}(\mathbf{B}) = -\rho_{ji}^{\mathrm{aS}}(\mathbf{B}) \tag{2–22}$$

Combining Eq. (19) with Eq. (21) gives

$$\rho_{ij}^{\mathrm{S}}(\mathbf{B}) = \rho_{ji}^{\mathrm{S}}(-\mathbf{B}) = \rho_{ij}^{\mathrm{S}}(-\mathbf{B}) \tag{2–23}$$

which states that ρ_{ij}^{S} is an *even* function of **B**. Similarly,

$$\rho_{ij}^{\mathrm{aS}}(\mathbf{B}) = -\rho_{ij}^{\mathrm{aS}}(-\mathbf{B}) \tag{2–24}$$

stating that ρ_{ij}^{aS} is an *odd* function of **B**. If these **B** dependences are then expanded in a power series to the first term in **B**, we obtain

$$\rho_{ij}^{\mathrm{S}}(\mathbf{B}) = \rho_{ij}^{\mathrm{o}} + \xi_{ijkl} B_k B_l \tag{2–25}$$

and

$$\rho_{ij}^{\mathrm{aS}}(\mathbf{B}) = R_{ijk} B_k \tag{2–26}$$

where ρ_{ij}^{o} is the ordinary resistivity tensor in the absence of magnetic induction. In this way we have introduced two new matter tensors, **R** and $\boldsymbol{\xi}$. The quantity $\boldsymbol{\rho}^{\mathrm{aS}}$ is an antisymmetric tensor of second rank and, therefore, can be regarded as an axial vector after converting to one-index notation in the usual way (see Appendix A):

ij	23	31	12
i	1	2	3

Thus Eq. (26) can be rewritten as

$$\rho_{i}^{\mathrm{aS}}(\mathbf{B}) = R_{ik} B_k \tag{2–27}$$

The tensor R_{ik}, which relates two axial vectors, is therefore a second-rank *polar* tensor; it is not symmetric, however. To see the meaning of its diagonal terms, consider R_{11} (which is R_{231} in three-index notation). This term refers to a magnetic induction perpendicular to an electric current flow giving rise to an electric field perpendicular to both. (See Fig. 2–1.)

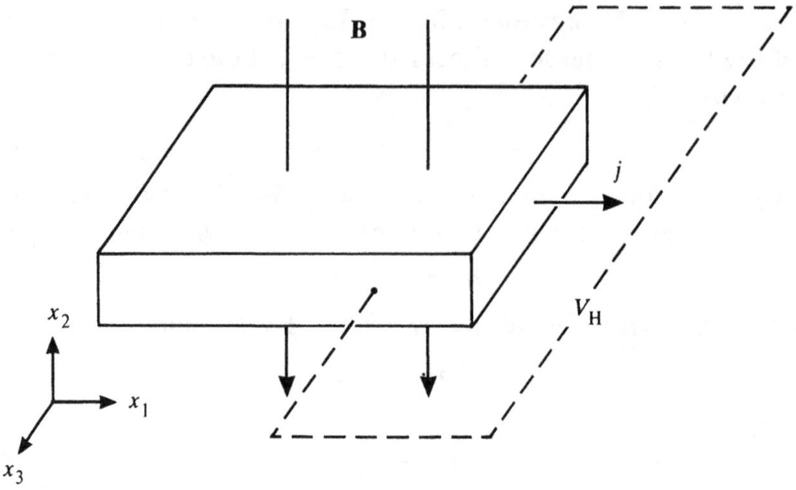

Fig. 2–1. The Hall effect: a current flow parallel to x_1 in the presence of magnetic induction parallel to x_2 gives rise to an electric field (or Hall voltage, V_H) parallel to x_3. Physically, this originates in the force on a moving charge given by $q(\mathbf{v} \times \mathbf{B})$, where q is the charge and $\mathbf{v}$ its velocity.

Such a field is called the 'Hall field' and the phenomenon is the well-known 'Hall effect'. Accordingly, R_{ik} is called the *Hall tensor*; it is the full tensor that constitutes a generalization of the Hall effect phenomenon.

Next we turn to the tensor ξ in Eq. (25), called the *magnetoresistivity tensor*. This property relates a symmetric tensor of second rank to products of axial vector components; therefore, it can be regarded as relating two symmetric tensors of second rank. The tensor ξ_{ijkl} is then of fourth rank, and is symmetric with respect to interchange of i and j and of k and l. Like the elastic compliance (Section 1–4), it can be converted to two-index notation; however, it is not symmetric in the interchange of i, j with k, l. It is instructive to examine the meaning of ξ_{1111} and ξ_{1122}. The former represents **B** applied parallel to direction 1 and gives the change ($\delta\rho_{11}$) of resistivity due to such a longitudinal magnetic field. The quantity ξ_{1122}, on the other hand, represents **B** applied in direction 2 and is the change in resistivity ($\delta\rho_{11}$) due to a transverse magnetic field. These two quantities are then called the longitudinal and transverse magnetoresistances, respectively. These two simple effects may be generalized into the complete magnetoresistivity tensor for the case of an arbitrary crystal and an arbitrary direction of the **B** field.

There are completely analogous effects of magnetic induction on the thermal conductivity tensor **K**. The *Righi–Leduc effect*, which is a transverse temperature gradient produced when **B** is applied perpendicular to a heat flux, is the analog of the Hall effect. Again, a complete tensor is defined as the Righi-Leduc tensor. Also the *magnetothermal resistivity* tensor is the exact analog of the magnetoresistance tensor.

The effect of a **B** field on the thermoelectric tensor $\mathbf{\Sigma}$ is somewhat different. Since $\mathbf{\Sigma}$, in the absence of magnetism, is not symmetric, its symmetric and antisymmetric parts are not even and odd functions of **B**, respectively, as we found for the resistivity, Eqs. (23) and (24). In expanding this tensor about $\mathbf{B} = 0$, we then write

$$\Sigma_{ij}(B) = \Sigma_{ij}^{o} + \Sigma_{ijk} B_k + \Sigma_{ijkl} B_k B_l + \ldots \qquad (2\text{–}28)$$

The matter tensor Σ_{ijk} (which is the change in the ith component of the electric field produced by the jth component of the temperature gradient due to the kth component of the **B** field) is analogous to the Hall tensor but does not have antisymmetry in i and j. It is called the *Nernst tensor* and is a generalization of the Nernst effect. That effect, which occurs even in an isotropic medium, is the appearance of a transverse potential difference when **B** is applied perpendicularly to a longitudinal thermal gradient (i.e. refer to Fig. 2–1, but now let **j** become a temperature gradient). Equation (28) shows that Σ_{ijk} relates a second-rank tensor to an axial vector. It is, therefore, a third-rank axial tensor. The converse effect is the *Ettinghausen tensor* obtained by expanding $\mathbf{\Sigma}'$ of Eq. (13) in **B**. This is a generalization of the Ettinghausen effect, which is the appearance of a temperature gradient perpendicular to both a longitudinal current flow and an applied magnetic field.

In view of the generalized Onsager relation, Eq. (18), the tensors $\mathbf{\Sigma}'$ and $\mathbf{\Sigma}$ are related by

$$\Sigma'_{ijk} = -\Sigma_{jik} \qquad (2\text{–}29)$$

The second-order effect Σ_{ijkl} in Eq. (28) relates a second-rank tensor to a second-rank symmetric tensor. Thus it is a fourth-rank tensor that is symmetric in k and l but not in i and j. It is called the *magnetothermoelectric power tensor*.

All the power series expansions of thermal and electric transport properties can be carried out further to third- and fourth-order effects in **B**. For the $\boldsymbol{\rho}$ and **K** tensors the symmetric and antisymmetric parts involve even and odd power series expansions, respectively, as in Eqs. (25) and (26). For the $\mathbf{\Sigma}$ tensor, all powers appear in a single equation, as in Eq. (28). For example, consider ρ_{ijklm}, the third-order effect of a **B** field on $\boldsymbol{\rho}$, obtained by carrying Eq. (26) to one more term, as follows:

$$\rho_{ij}^{aS}(\mathbf{B}) = R_{ijk} B_k + \rho_{ijklm} B_k B_l B_m \qquad (2\text{–}30)$$

This effect is often regarded as a *second-order Hall effect*. It is antisymmetric in i and j and totally symmetric in the indices k, l and m. It relates an antisymmetric tensor of second rank (or an axial vector) to a totally symmetric third-rank axial tensor. As in the case of the Hall effect tensor, it is advantageous to reduce the indices i, j of ρ_{ij}^{aS} to a single index, thus leaving our second-order Hall effect tensor in a four-index notation, as ρ_{iklm}.

2–5 Summary of transport properties

In summary, most of the remarks made at the end of Chapter 1 (Section 1–8) apply as well to the transport properties as to the equilibrium properties. In Table 2–1 we summarize the relevant features of all the transport properties that have been discussed in this chapter including the tensor character, the types of physical quantity being related, the intrinsic symmetries and the maximum number of constants that make up the tensor in question. Again, we postpone detailed discussion of specific materials that show these properties until we have explored the effects of crystal symmetry, in Chapters 6–10.

We will see later (in Chapter 5) that the question of 'special magnetic properties' does not arise in connection with transport properties.

2–6 Properties that cannot be represented by tensors

This section contains some concluding remarks applicable to both Chapters 1 and 2. Thoughout these chapters we have defined properties representing a unique relationship between a 'force' and a 'response'. There are many cases in which a given 'force' does not produce a unique response, as when hysteresis or time-dependent phenomena take place. The relevant properties cannot be dealt with by the methods of this book.

For example, in ferromagnetic materials subject to an external magnetic field the magnetization is a function of previous history; under an alternating field a hysteresis loop is produced in the B vs H curve. Similarly, ferroelectric materials subjected to an alternating electric field show hysteretic D vs. E behavior. Another example is that of a material subjected to a stress sufficient to produce plastic deformation. Such mechanical behavior is irreversible and thus cannot be described in terms of a unique stress–strain relationship. Finally, time-dependent processes, such as creep (time-dependent deformation), also fall outside the domain of behavior that can be described by matter tensors.

Table 2–1. *Transport properties*

Property	Tensor character	Relation between	Symbol	Symmetric with respect to:	Maximum no. of constants
Conductivity (resistivity) Thermal conductivity Diffusivity	$T_S(2)$	$T(1)$ and $T(1)$	σ_{ij} (ρ_{ij}) K_{ij} D_{ij}	$i \longleftrightarrow j$	6
Thermoelectric power	$T(2)$	$T(1)$ and $T(1)$	Σ_{ij}	–	9
Hall effect Righi–Leduc	$T(2)$	$T(1)^{ax}$ and $T(1)^{ax}$	R_{ij} –	–	9
Nernst	$T(3)^{ax}$	$T(1)^{ax}$ and $T(2)$	Σ_{ijk}	–	27
Magnetoresistance Magnetothermal resistivity Piezoresistance	$T(4)$	$T_S(2)$ and $T_S(2)$	ξ_{ijkl} – Π_{ijkl}	$i \longleftrightarrow j,\ k \longleftrightarrow l$	36
Magnetothermoelectric power	$T(4)$	$T_S(2)$ and $T(2)$	Σ_{ijkl}	$k \longleftrightarrow l$	54
Second-order Hall	$T(4)$	$T(1)^{ax}$ and $T_S(3)^{ax}$	ρ_{iklm}	$k \longleftrightarrow l \longleftrightarrow m$	30

Problems

2–1. Carry out the conversion of Eqs. (10) into Eqs. (13).

2–2. Consider the meaning of an arbitrary coefficient ξ_{ijkl} of the magnetoresistivity tensor.

2–3. Do the same for the component R_{ik} of the Hall tensor.

2–4. Write the equation (analogous to Eq. (28)) that defines the Ettinghaussen tensor Σ'_{ijk} and show that Eq. (29) applies. Do the equivalent for the tensor Σ'_{ijkl}.

3

Review of group theory

A crystalline solid is characterized by the fact that it possesses spatial periodicity of its atomic arrangement. It is usual to view the basic motif as a *unit cell* that repeats itself in each of three non-planar directions in three-dimensional space (constituting the 'translational symmetry' of the crystal). Real crystals are never perfect, however, and many of their properties depend on the various point, line and surface defects that can be present. Nevertheless, the perfect crystal remains an idealized reference state against which the behavior of real crystals must be compared.

3–1 Crystal symmetry and the point groups

An important characteristic of a crystal is the symmetry that it possesses in its atomic arrangement. A *symmetry operation* is one which takes the crystal from its initial configuration into an equivalent (indistinguishable) configuration. The symmetry operations include the following:*

- Reflection across a plane of symmetry, or mirror plane, denoted by σ.
- Inversion through a center of symmetry, denoted by i. Under inversion, every point (x, y, z) is taken into $(-x, -y, -z)$.
- Rotation about an axis of symmetry denoted by C_n^m which means rotation by $(2\pi/n)m$, where $n = 2$, 3, 4 or 6 and m is an integer $<n$. Thus C_3 is a rotation through 120° and C_3^2 through 240°.
- Improper rotation S_n^m, implying a rotation C_n^m preceded or followed by reflection through a plane perpendicular to the rotation axis.

We illustrate some of the above symmetry operations by considering a

* We consider here only point symmetry operations of crystals, and not their translational symmetry, which leads to the space groups. This further implies that if a crystal has the symmetry of a glide plane or a screw axis (which, respectively, combine a reflection or a rotation with a translational motion), the translational part is omitted from consideration.

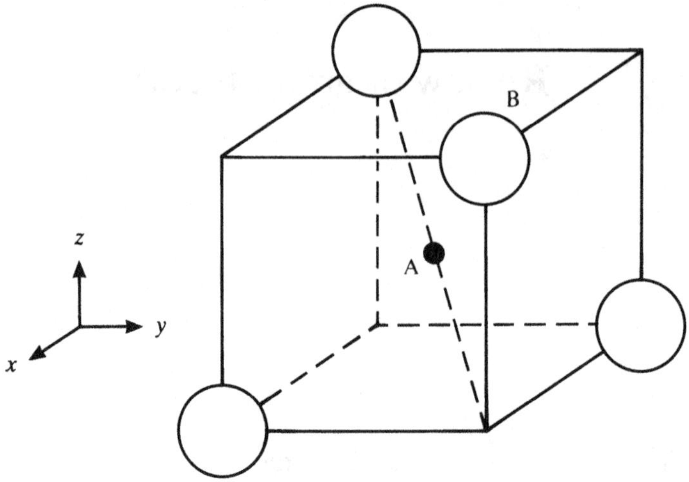

Fig. 3–1. A tetrahedral molecule of the type AB_4.

tetrahedral molecule of the type AB_4 (e.g. CH_4), as shown in Fig. 3–1. The reader may easily verify that, for this molecule, with the axes shown, all six {110} planes are mirror planes, that the four <111> axes are three-fold axes of symmetry (C_3), and that the three cube, or <100>, axes are four-fold axes of improper rotation (S_4). (Note that we refer here to planes and directions by the well-known Miller indices, in spite of the fact that this is only a single molecule.)

We also use the term *symmetry elements* referring to the geometric entities with respect to which the symmetry operations are performed, for example the plane, the center of inversion and the axis of rotation or of rotation–reflection.

By a product of two symmetry operations AB, we mean that which occurs if we first carry out operation B and then follow this by A. It is then possible to set up a multiplication table of the symmetry operations. It is easily shown (see the standard texts) that the set of all symmetry operations constitutes a group, that is, that they obey the four group postulates which are:

- They include the *identity* operation E (i.e. leaving the crystal as is).
- The set is *closed*, in that the product of any two symmetry operations is a third symmetry operation.
- Every operation A has an *inverse* A^{-1} (defined so that the product AA^{-1} or $A^{-1}A$ is the identity E.
- The association rule is obeyed, that is, $A(BC) = (AB)C$.

Particularly noteworthy in the defining postulates is the absence of a commutation rule, that is, $AB = BA$. Some operations will, of course, commute. In particular, the identity E commutes with every operation of the group. If a group does have the property that *all* its elements commute with each other, it is said to be an *Abelian* group. A large fraction of the point groups, however, have some non-commuting operations and are said to be *non-Abelian*.

An important example that illustrates some of the features of groups is the smallest non-Abelian group, containing six members, which we designate in abstract form as E, A, B, C, D and F (where E is the identity), and which obey the following multiplication table:

	E	A	B	C	D	F
E	E	A	B	C	D	F
A	A	B	E	F	C	D
B	B	E	A	D	F	C
C	C	D	F	E	A	B
D	D	F	C	B	E	A
F	F	C	D	A	B	E

Note that this group is non-Abelian (since CD ≠ DC, DF ≠ FD, etc.). We will see later that two of the crystallographic point groups obey this multiplication table.

Within a group, the concept of *classes* of operations is very important. Operations A and B are said to be in the same class if there exists a third operation C in the group that converts A into B or vice versa. Formally, this can be expressed as $C^{-1}AC = B$. Clearly, A and B, must be equivalent operations (e.g. both mirror planes, or both 90° rotations). Note that in an Abelian group every operation is in a class by itself. Another important concept is that of the *generators* of a group. These are the minimum collection of operations which, if applied over and over, singly and in combination (if there is more than one), will produce all the operations of the group. A group that can be produced by the repeated application of a single generator is called a *cyclic group*. Such a group is always Abelian. (Why?) For example, the cyclic group with generator C_4 contains the operations C_4, C_2, C_4^3 and E. (This group is named the C_4 group.)

The non-Abelian group of order six whose multiplication table is given above provides good examples of classes and generators. For this group operations A and B are in the same class, and so are the three operations

C, D and F. On the other hand, identity E is always in a class by itself. For generators of this group, we may take A and C, since repeated application of these two operations, separately and in combination, will generate the entire group.

Also of importance is the concept of a *subgroup*, which is a subset of the original group that itself constitutes a group. It can be shown that the order (number of members) of a subgroup must be a submultiple of the order of the original group. Thus, for the non-Abelian group of order 6, aside from the trivial case of a group consisting only of the identity, possible subgroups can only be of order 2 or 3. The reader may verify that the subsets E, A, B and also E, D constitute subgroups of this group.

For crystals, there exist only 32 possible combinations of symmetry operations that obey the above considerations. These are the 32 classical crystallographic *point groups* which are also often termed the 32 crystal classes (not to be confused with 'classes' as applied to the group operations). Figure 3–2 presents a pair of diagrams for each of the crystallographic point groups. These diagrams are sterographic projections of a sphere, in which the axes x and y lie in the page, and the axis z is perpendicular to it. Whenever there is a single preferred axis, as in the case of uniaxial crystals with a three-fold, four-fold or six-fold axis, the convention is to take the preferred axis along the z-direction. The x- and y-axes are chosen in the conventional manner. The second diagram in each case shows the symmetry elements of the group, a reflection plane being shown as a heavy line or circle, a two-fold axis of rotation by an ellipse, a three-fold axis as a triangle, a four-fold axis as a square and a six-fold axis as a hexagon, all solid (i.e. filled in). In addition, open symbols are used for the product of the appropriate rotation by an inversion, i. Thus, open triangles, squares or hexagons appear for some of the groups. The first diagram for each group shows all the points into which an arbitrary point in the projection is taken by the symmetry operations of the group. Thus, the number of points is equal to the number of symmetry operations or the *order of the group*, h. In these diagrams a dot may be regarded as located in the position shown in the x–y-plane and an arbitrary distance *above* the plane (i.e. in the z-direction), a circle is located at the x–y-position shown, but an equal distance *below* the plane (i.e. in the $-z$-direction). The reader who has no previous experience with such diagrams should verify the first diagram in terms of the second for a few of the groups, for example for the group D_{2d}. In such an exercise, it is convenient to take a point in the first quadrant as a reference point. The property of closure of the group manifests itself by the fact that any group operation carried out on any of the points shown will lead to another such point.

For molecular symmetry, there exist other point groups besides the 32 given here. These include groups with five-fold and eight-fold axes of rotation, for example. In the case of crystals, however, because of the additional requirement of translational symmetry in three dimensions, only groups with two-, three-, four- and six-fold axes are possible.*

The nomenclature for the point groups requires mention. Two sets of names are used: the Schoenflies notation (e.g. C_{4v}) and the International (or Hermann–Mauguin) notation (e.g. 4 mm). The former notation is mostly used in the literature of molecular studies and in work on defects in crystals, while the latter is favored by crystallographers involved in diffraction studies. In this book, we will primarily use the Schoenflies notation. Here C_n indicates the presence of an n-fold symmetry axis, while the subscripts h and v refer to 'horizontal' and 'vertical' mirror planes, which will be defined below. The symbol D_n refers to a group which, in addition to the principal n-fold axis along the z-direction, also has two-fold axes lying in a prependicular plane (i.e. in the $x-y$-plane). The symbol S_n refers to an n-fold improper rotation axis. The cubic groups have a notation of their own, T standing for 'tetrahedral' and O for 'octahedral' symmetry.

In the International notation, a first number 6 represents a six-fold axis, $\bar{6}$, a six-fold axis multiplied by an inversion, while subsequent m's refer to mirror planes and 2's to perpendicular two-fold axes. Finally, n/m refers to a mirror plane perpendicular to the principal (n-fold) axis.

Table 3–1 lists all the groups with their symmetry operations and generating elements. In this listing, the following standard terminology is used for mirror planes. A plane perpendicular to the major symmetry axis (z-axis) is called a 'horizontal' plane and is denoted by σ_h. A plane containing the symmetry axis is called 'vertical' and is denoted by σ_v. Additional vertical mirror planes which lie in directions that bisect the principal vertical planes are denoted by σ_d. Thus, for example, in the group D_{4h}, σ_h is the $x-y$-plane, the two σ_v are the yz- and xz-planes, while the two σ_d planes are parallel to the z-axis and at angles 45° and 135° to the $x-z$-plane. In the case of two-fold axes of rotation, the symbol C_2 is used if there is only one such axis. But in cases where there are more than

* Recently, structures showing five-fold symmetry in the form of sharp diffraction spots have been observed and dubbed 'quasicrystals'. Much has been written about such structures (for a review, see Janot, 1992), which can be viewed as involving more than one type of cell stacked together so as to fill all space. Such structures, however, fall outside the range of interest of the present book.

Triclinic	Monoclinic (1st setting)	Tetragonal
C_1 1	C_2 2	C_4 4
—	C_s $m(-\bar{2})$	S_4 $\bar{4}$
C_i $\bar{1}$	C_{2v} $2/m$	C_{4h} $4/m$
Monoclinic (2nd setting) C_2 2	**Orthorhombic** D_2 222	D_4 422
C_s m	C_{2v} $mm2$	C_{4v} $4mm$
—	—	D_{2d} $\bar{4}2m$
C_{2v} $2/m$	D_{2h} mmm	D_{4h} $4/mmm$

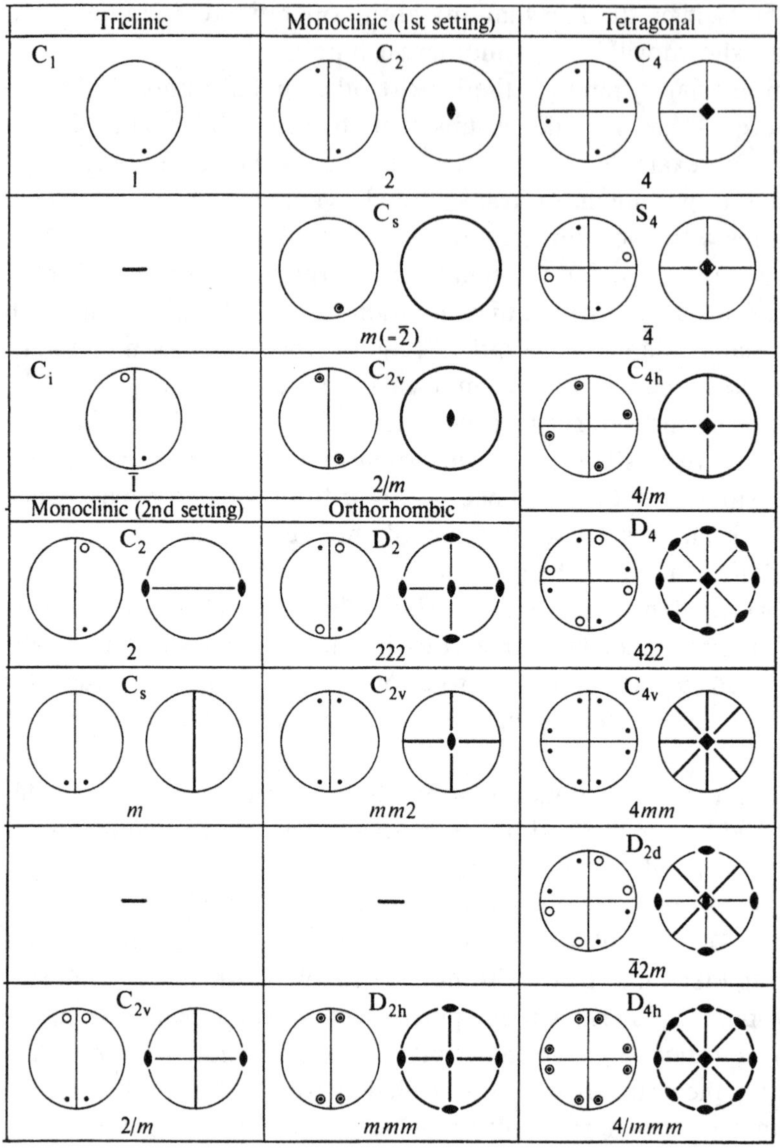

one kind, C_2 is used for the principal axis and C_2' or C_2'' for the secondary axes. Thus, in group D_4, C_2 denotes the z-axis rotation ($= C_4^2$) while C_2' and C_2'' denote the two-fold axes of rotation that lie perpendicular to the z-axis. By listing symmetry operations according to classes, the number of entries in Table 3–1 can be reduced. The reader should recall that two

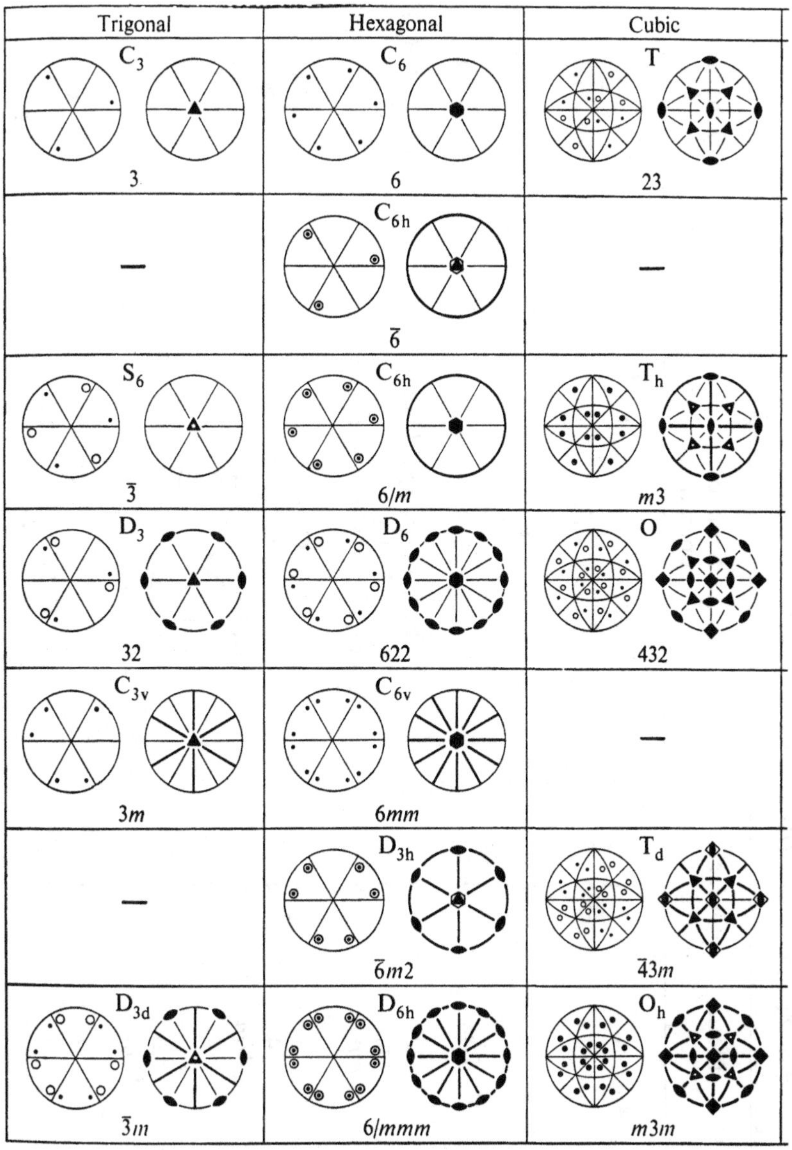

Fig. 3–2. Stereograms of the 32 point groups, listed by crystal system. (From the *International Tables for X-ray Crystallography*.) For each group, the upper name is that of the Schoenflies and the lower, that of the International notation.

operations are in the same class if there exists a third operation that takes one into the other. Thus, for example, in the group C_{4v}, the two rotations C_4 and C_4^3 can be interchanged by one of the vertical mirror planes; therefore, they are in the same class. Similarly, the two vertical planes σ_v are taken into each other by a C_4 operation and are therefore in the same

Table 3–1. *The 32 crystallographic point groups listed by crystal system. For each group, the nomenclatures, order, symmetry operations (listed by class), and generators are given*

	Schoenflies	Intern.	Order (h)	Symmetry operations	Generators
Triclinic					
	C_1	1	1	E	E
	C_i	$\bar{1}$	2	E, i	i
Monoclinic					
	C_2	2	2	E, C_2	C_2
	C_s	m	2	E, σ	σ
	C_{2h}	$2/m$	4	E, C_2, i, σ_h	i, C_2
Orthorhombic					
	D_2	222	4	$E, C_2(x), C_2(y), C_2(z)$	$C_2(y), C_2(z)$
	C_{2v}	$mm2$	4	$E, C_2, \sigma_v, \sigma'_v$	C_2, σ_v
	D_{2h}	mmm	8	$E, C_2(x), C_2(y), C_2(z), i, \sigma^{xy}, \sigma^{xz}, \sigma^{yz}$	$i, C_2(y), C_2(z)$
Tetragonal					
	C_4	4	4	E, C_4, C_2, C_4^3	C_4
	S_4	$\bar{4}$	4	E, S_4, C_2, S_4^3	S_4
	C_{4h}	$4/m$	8	$E, C_4, C_2, C_4^3, i, S_4, \sigma_h, S_4^3$	C_4, i
	D_4	422	8	$E, 2C_4, C_2, 2C'_2, 2C''_2$	C_4, C'_2
	C_{4v}	$4/mm$	8	$E, 2C_4, C_2, 2\sigma_v, 2\sigma_d$	C_4, σ_v
	D_{2d}	$\bar{4}2m$	8	$E, 2S_4, C_2, 2C'_2, 2\sigma_d$	S_4, C'_2
	D_{4h}	$4/mmm$	16	$E, 2C_4, C_2, 2C'_2, 2C''_2, i, 2S_4, \sigma_h, 2\sigma_h, 2\sigma_d$	C_4, i, C'_2
Trigonal					
	C_3	3	3	E, C_3, C_3^2	C_3
	S_6	$\bar{3}$	6	$E, S_6, C_3, i, C_3^2, S_6^5$	S_6
	D_3	32	6	$E, 2C_3, 3C_2$	C_3, C_2
	C_{3v}	$3m$	6	$E, 2C_3, 3\sigma_v$	C_3, σ_v
	D_{3d}	$\bar{3}m$	12	$E, 2C_3, 3C_2, i, 2S_6, 3\sigma_d$	C_3, C_2, i

Table 3–1. *(cont.)*

	Schoenflies	Intern.	Order (h)	Symmetry operations	Generators
Hexagonal					
	C_6	6	6	$E, C_6, C_3, C_2, C_3^2, C_6^5$	C_6
	C_{3h}	$\bar{6}$	6	$E, S_3, C_3^2, \sigma_h, C_3, S_3^5$	S_3
	C_{6h}	$6/m$	12	$E, C_6, C_3, C_2, C_3^2, C_6^5, i, S_3^5, S_6^5, \sigma_h, S_6, S_3$	C_6, i
	D_6	622	12	$E, 2C_6, 2C_3, C_2, 3C_2', 3C_2''$	C_6, C_2'
	C_{6v}	$6mm$	12	$E, 2C_6, 2C_3, C_2, 3\sigma_v, 3\sigma_d$	C_6, σ_v
	D_{3h}	$\bar{6}m2$	12	$E, 2C_3, 3C_2, \sigma_h, 2S_3, 3\sigma_v$	S_3, C_2
	D_{6h}	$6/mmm$	24	$E, 2C_6, 2C_3, C_2, 3C_2', 3C_2'', i, 2S_3, 2S_6, \sigma_h, 3\sigma_d, 3\sigma_v$	C_6, C_2', i
Cubic					
	T	23	12	$E, 4C_3, 4C_3^2, 3C_2$	$C_3[111], C_2$
	T_h	$m3$	24	$E, 4C_3, 4C_3^2, 3C_2, i, 4S_6, 4S_6^5, 3\sigma_h$	$C_3[111], C_2, i$
	O	432	24	$E, 8C_3, 6C_4, 3C_2, 6C_2'$	$C_3[111], C_4$
	T_d	$\bar{4}3m$	24	$E, 8C_3, 3C_2, 6S_4, 6\sigma_d$	$C_3[111], S_4$
	O_h	$m3m$	48	$E, 8C_3, 6C_4, 3C_2, 6C_2', i, 8S_6, 6S_4, 3\sigma_h, 6\sigma_d$	$C_3[111], C_4, i$

class. In the group C_4, however, rotation C_4 and rotation C_4^3 are each in a class by itself. In fact, it can easily be shown that in any Abelian group, and, in particular, any cyclic group, every operation is in a class by itself.

The reader should note that the crystallographic groups C_{3v} and D_3 both obey the same multiplication table as the first non-Abelian abstract group (page 33), and have the same class structure as this group. (See Problem 3–5.)

The groups in Table 3–1 and Fig. 3–2 are listed according to the seven *crystal systems*. The lowest crystal system is the *triclinic* system which possesses at most, only a center of symmetry. Next, the *monoclinic* system has one two-fold axis or a plane of symmetry, while the *orthorhombic* has three two-fold axes and/or symmetry planes that are mutually perpendicular. The uniaxial systems each have a major axis of symmetry (taken as the z-axis by convention). These include the *trigonal* (three-fold axis), *tetragonal* (four-fold axis), and *hexagonal* (six-fold axis) systems.*

Finally, *cubic* crystals possess four three-fold axes located along the diagonals of a cube, so that the x-, y- and z-axes (chosen to be along the cube edges) are taken into each other by these operations. Thus, one of the generators of the group is always a C_3 [111]. Note that the cubic groups do not necessarily have C_4 axes along the <100> directions (only the groups O and O_h have such four-fold axes).

3–2 Representation theory

In this section we introduce the very important concept that a group can be 'represented' by a set of matrices. Throughout this section we will be quoting important theorems of this representation theory as required; proofs may be found in standard textbooks on group theory (see the references).

An n-dimensional *vector space* is formed by a set of n quantities $\mathbf{e}_i$, called *basis vectors*, which are linearly independent and where the most general vector quantity in the same space may be written as a linear combination of the $\mathbf{e}_i$'s:

$$\mathbf{v} = x_i\mathbf{e}_i \qquad (3\text{–}1)$$

Here the quantities x_i are called the *components* of the hypervector $\mathbf{v}$. One can define a scalar product in the usual way, and with this definition

* It is customary to consider a group such as C_{3h} or D_{3h} as hexagonal (rather than trigonal) since it contains the product iC_6 (called $\bar{6}$ in international notation); similarly the group S_6, which contains iC_3, is regarded as trigonal.

choose the basis vectors $\mathbf{e}_i$ of the space to be orthogonal and normalized, without loss of generality.

Now consider the effect of a symmetry operation, R, on the vector $\mathbf{v}$. We may write

$$R\mathbf{v} = \mathbf{v}' \tag{3–2}$$

that is, the operation R takes $\mathbf{v}$ into a new vector $\mathbf{v}'$. If we keep the basis vectors $\mathbf{e}_i$ fixed, the components x_i will be transformed into a new set x_i' given by

$$x_i' = [D(R)]_{ij}x_j \tag{3–3}$$

that is, the new components are linear combinations of the original ones. The set of coefficients $[D(R)]_{ij}$ form an $n \times n$ matrix which is related to the operation R. By way of illustration, consider the case in which the basis vectors $\mathbf{e}_i$ are just the unit vectors in three-dimensional space, so that the x_i are the components x, y and z. If R is a rotation through angle θ about the z-axis, the matrix $\mathbf{D}(R)$ is then given by

$$\mathbf{D}(R) = \begin{pmatrix} \cos\theta & -\sin\theta & 0 \\ \sin\theta & \cos\theta & 0 \\ 0 & 0 & 1 \end{pmatrix}$$

The entire set of operations $\{R\}$ of the point group of the crystal generates a set of such matrices called a *representation* Γ of the group (henceforth, to be abbreviated as a 'rep'). It is customary to say that the vector $\mathbf{v}$ is the basis of rep Γ. It can be shown that the matrices of such a rep obey the same multiplication table (for matrix multiplication) as that of the group itself. Thus, if A, B and C are group operations such that $AB = C$, then

$$[D(A)]_{ij}[D(B)]_{jk} = [D(C)]_{ik} \tag{3–4}$$

It is convenient (and always possible) to confine ourselves to reps composed of *unitary matrices*, so that

$$[\mathbf{D}(R)]^{-1} = \mathbf{D}(R)^{\dagger} \tag{3–5}$$

where the symbol $\dagger$ means the transposed complex conjugate. (If the matrices are real, $\mathbf{D}(R)$ is then an orthogonal matrix, but to handle all the crystallographic point groups it is necessary to allow for complex reps.)

Next we wish to consider the effect of a change of axes, that is, of basis vectors, on a rep $\mathbf{D}(R)$. A change of basis vectors can be written as

$$\mathbf{e}' = \mathbf{eS} \tag{3–6}$$

where the individual vectors $\mathbf{e}_i$ or $\mathbf{e}_i'$ are here arranged in rows (e.g. $\mathbf{e} = (\mathbf{e}_1, \mathbf{e}_2, \ldots)$) and $\mathbf{S}$ is the transformation matrix conveniently taken as unitary. It can then be shown (see Problem 3–7) that such a change of axes changes the rep matrices to

$$\mathbf{D}'(R) = \mathbf{S}^{-1}\mathbf{D}(R)\mathbf{S} \tag{3–7}$$

which is called a 'similarity transformation'. Reps $\Gamma' = \{\mathbf{D}'(R)\}$ and $\Gamma = \{\mathbf{D}(R)\}$ are then said to be *equivalent* to each other, since they only differ in the choice of axes.

An important property of the matrix $\mathbf{D}(R)$ of a rep is its trace or diagonal sum $\chi(R)$, called its *character*:

$$\chi(R) = D_{ii}(R) \tag{3–8}$$

For example, for the case of a rotation through an angle θ about the z-axis in three-dimensional space, $\chi(R) = 1 + 2\cos\theta$. Also, for the identity operation E, $D_{ij}(E) = \delta_{ij}$, where δ_{ij} is the Kronecker delta, and so $\chi(E) = n$, where n is the dimension of the rep. It is straightforward to show that the character of a matrix does not change under a similarity transformation. There are two important consequences of this theorem. The first is that the set of characters of a rep remains the same independently of the choice of axes (i.e. of the system of basis vectors). The second is that all matrices representing members of the same class of a given group have the same character, that is, that, while $D_{ij}(R)$ and $D_{ij}(S)$ are different, $\chi(R)$ and $\chi(S)$ are the same if R and S belong to the same class.

An important aspect of a rep is its reducibility. A rep Γ, consisting of the set of matrices $\mathbf{D}(R)$, is said to be *reducible* if there exists a similarity transformation, Eq. (7), that converts *all* the matrices of the rep into the same block form. For example, they may all be taken into a block pair that looks as follows:

$$\mathbf{D}'(R) = \left(\begin{array}{c|c} \mathbf{D}_1(R) & \mathbf{0} \\ \hline \mathbf{0} & \mathbf{D}_2(R) \end{array}\right) \tag{3–9}$$

Here $\mathbf{D}_1(R)$ is an $n_1 \times n_1$ matrix and $\mathbf{D}_2(R)$ an $n_2 \times n_2$ matrix, with $n_1 + n_2 = n$. (Clearly the zeros are also matrices.) If a rep cannot be so reduced it is said to be an *irreducible representation*. For brevity, the latter will be referred to as an *irrep* throughout this book. The importance of the question of reducibility is as follows. Suppose that a rep is reducible. This means that there exist subspaces of the original vector space such that, with proper choice of basis vectors, all the group operations take only the vector components in each subspace into linear combinations of each other; there is no mixing of components belonging to different subspaces. For example, imagine that we are dealing with the three-dimensional x, y,

z-space, where the set of three-dimensional matrices of a rep is reduced to blocks of two for the coordinates x and y and of one for the coordinate z. In this case the group operations take x and y only into linear combinations of x and y, while z is always transformed only into itself (multiplied by a constant); thus, coordinates x and y are completely decoupled from z. (We then say that x and y belong to a two-dimensional irrep while z belongs to a one-dimensional irrep.) Such decoupling provides a considerable simplification of the problem of finding the effects of symmetry, all the more so if the number of coordinates, n, is relatively large, for example 9 or 12.

One of the basic questions of representation theory is to determine, for a given point group, how many independent irreps there are, and what forms they take. Part of the answer comes from the following theorems:

1. The number of irreps of a given group is equal to the number of classes of its symmetry operations.
2. If h is the order of the group and l_α the dimensionality of the αth irrep, then

$$\Sigma_\alpha l_\alpha^2 = h \tag{3–10}$$

where the sum is taken over all irreps.

Another important problem is: given a reducible rep Γ, to decompose it into irreps. (In general, the possibility is not excluded that a given irrep may appear more than once in such a decomposition.) It is clear from the block form (such as that of Eq. (9)), that the character of each matrix in the rep will be the sum of the characters of the various irreps that appear, or

$$\chi(R) = \Sigma_\gamma m_\gamma \chi^{(\gamma)}(R) \tag{3–11}$$

where m_γ is the number of times that irrep γ appears in the decomposition of Γ, $\chi(R)$ is the character of the matrix $D(R)$ of Γ, and $\chi^{(\gamma)}(R)$ is the character of irrep γ of the group for operation R. In order to carry out this decomposition, we have the *decomposition theorem*, which states that the number of times, m_α, that irrep α appears in the decomposition of Γ is given by

$$m_\alpha = (1/h)\Sigma_R \chi^{(\alpha)}(R)^* \chi(R) \tag{3–12}$$

where the asterisk denotes the complex conjugate. In other words, if we know the characters of the irreps as well as those of a given rep Γ, we can obtain the decomposition of rep Γ. It is therefore useful to develop character tables which list all the irreps of a point group and give their characters.

3–3 The character tables

The character tables for the irreps of each crystallographic group may be obtained as follows. First the number of irreps and their dimensionalities l_α are obtained from the two theorems quoted above, which include Eq. (10). To these theorems we add the following:

3. Existence of a *totally symmetric irrep*: every group has a one-dimensional irrep for which $\chi(R) = D(R) = 1$ for every operation R. (Why is there always such an irrep?) This irrep is usually listed first in the character table and will often be referred to as the A_1 irrep (although it may have other names, e.g. A, A′, A_g or A_{1g}).
4. Since $\mathbf{D}(E)$ for the identity operation must be the (diagonal) unit matrix of dimensionality l_α, it follows that $\chi^{(\alpha)}(E) = l_\alpha$.
5. The orthogonality theorem for characters: If α and β are two irreps of a group $\mathcal{G}$ of order h, then

$$\Sigma_R \chi^{(\alpha)}(R)^* \chi^{(\beta)}(R) = h\delta_{\alpha\beta} \tag{3–13}$$

   where the summation is taken over all the group operations, R, and $\delta_{\alpha\beta}$ is the Kronecker delta. This theorem follows directly from the 'Great Orthogonality Theorem', which is the heart of group representation theory. (See Appendix D.)

It is also useful to note that, for a one-dimensional irrep, $\chi^{(\alpha)}(RS) = \chi^{(\alpha)}(R)\,\chi^{(\alpha)}(S)$. The reason is just that, for the one-dimensional case, the characters are the same as the matrices $D^{(\alpha)}(R)$, and that these matrices obey the same multiplication table as the group operations. (See Eq. (4).)

With these principles it is possible to construct the character tables. In these tables, the symmetry operations are listed across the top (by classes, recalling that all members of a class have the same χ), while the different irreps are listed vertically. Then in view of Theorem (1) above, the table is a square array. In the special case of Abelian groups, where every operation is in a class by itself, all irreps are one-dimensional. The complete set of character tables for all the cystallographic point groups may be found in Table 3–2 at the end of this chapter, page 55. In these tables we have used the well-known Mulliken notation for the irreps. In this nomenclature, one-dimensional irreps are designated by the letters A or B, two-dimensional irreps by E and three-dimensional irreps (which occur for cubic groups only) by T. Additional subscripts or primes are added as necessary, that is, when there is more than one irrep of that letter designation.

For a better understanding of many of the character tables of groups of higher order, it is helpful to be aware of the following considerations. Suppose we start with a group $\mathscr{G}_1$ that does not contain inversion symmetry i. We may then create a new group $\mathscr{G}_2$ with double the order of $\mathscr{G}_1$ by introducing the operation i. This larger group may be written as

$$\mathscr{G}_2 = \mathscr{G}_1 \times \mathrm{i}$$

which implies that it includes all members of $\mathscr{G}_1$ both in their original form and multiplied by i. Since i commutes with every element of $\mathscr{G}_1$, the new group has the property that it retains the class structure of $\mathscr{G}_1$ and adds a matching set of classes given by $\{\mathrm{i}R\}$ where $\{R\}$ is a class in $\mathscr{G}_1$. Then the character table of the new group $\mathscr{G}_2$ may be written in the block form:

	Classes of $\mathscr{G}_1$	New classes
(old irreps)$_g$	(+)	(+)
(old irreps)$_u$	(+)	(−)

where the symbol (+) implies all the characters of group $\mathscr{G}_1$, and (−) implies these same characters all multiplied by −1. Thus, both the class structure and the irreps are doubled, half the irreps now being designated sub g (for 'gerade' or *even* with respect to inversion symmetry) and the other half, sub u (for 'ungerade' or *odd*). The former do not change the sign of their characters under inversion (also referred to as having 'even parity') and the latter reverse their signs under this operation (having 'odd parity'). It is readily seen that the character table given above meets the requirement of the orthogonality theorem for characters, Eq. (13). Examples of such enlarged groups are: $C_{nh} = C_n \times \mathrm{i}$ (for $n = 2, 4, 6$), $D_{nh} = D_n \times \mathrm{i}$ ($n = 2, 4, 6$), $D_{3d} = D_3 \times i$, $T_h = T \times i$, and $O_h = O \times \mathrm{i}$. (The reader should check the character tables for some of these examples.) A similar possibility exists for the introduction of a horizontal mirror plane to produce groups $\mathscr{G}_2 = \mathscr{G}_1 \times \sigma_h$. In this case, the two sets of irreps are denoted by adding primes and double primes to the original irreps of $\mathscr{G}_1$. Examples are $C_{3h} = C_3 \times \sigma_h$ and $D_{3h} = D_3 \times \sigma_h$.

Another matter that requires special attention is the case of complex characters. In certain low-symmetry groups complex numbers appear among the characters of some one-dimensional irreps. (Such complex quantities are always one of the nth roots of unity, where an operation of the type C_n or S_n is involved.) In such cases, the irreps appear as pairs in which the characters are complex conjugates of each other. It is customary to list these pairs together and to label them with a designation $\tilde{E}$. These pairs of irreps have certain features in common with two-dimensional irreps, but it is important to remember that they are not two-dimensional; the tilde over the E serves as a reminder of that fact. The group C_6 provides a good example of a group that possesses complex characters.

In the conventional character tables that appear in standard textbooks on group theory, two additional columns are attached to the tables of the characters themselves. These extra columns are given in Tables 3–2 only for three groups: C_6, C_{4v} and D_{3d}, since they will be needed for discussion in the next section. For the remaining groups, these columns are omitted, and replaced by the separate Symmetry-Coordinate Transformation Tables, which will be described in Chapter 4.

Included among the character tables of Table 3–2 are two additional ones that represent infinite groups that are not members of the set of the 32 crystallographic point groups. These are the groups $C_{\infty v}$ and R(3). The group $C_{\infty v}$, which is the limiting case of C_{nv} as $n \to \infty$, represents the symmetry of a cylinder that has polarity. Its symmetry operations include all rotations (by angle ϕ) about the z-axis as well as all mirror planes containing the z-axis. The full rotation group has, for its symmetry operations, all the rotations of a sphere. These groups will be useful as limiting cases of polycrystalline materials: $C_{\infty v}$ to represent, for example, a poled polycrystalline ceramic in which dipole moments are preferentially aligned in the z-direction, R(3) to represent an isotropic material (e.g. a glass or a fully random polycrystal). Since these two groups have an infinite number of classes of symmetry operations, they have an infinite number of irreps as well (but a denumerable infinity).

3–4 Concept of symmetry coordinates

Consider an n-dimensional hypervector $\mathbf{X}$ (with components X_i) as the basis of an n-dimensional rep Γ of a group, in the manner of Eqs. (2) and (3). Then, by the decomposition formula, Eq. (12), we can determine how many times each irrep appears in Γ, that is, $\Gamma = m_\gamma \Gamma_\gamma$, where Γ_γ is the set of matrices for irrep γ, and the summation over irreps γ means that the matrices are placed in block form. But in terms of the original coordinates X_i which compose the hypervector, the rep Γ will, in general, not be in block form. It only goes into block form after an appropriate similarity transformation, Eq. (7). At the same time the original hypervector $\mathbf{X}$ goes into a new vector $\bar{\mathbf{X}}$ given by

$$\bar{\mathbf{X}} = \mathbf{SX} \qquad (3\text{–}14)$$

In component form we may write

$$\bar{X}_k = S_{ki}X_i \qquad (3\text{–}15)$$

However, because the coordinates $\bar{X}_k$ are associated with the irreps, the index k is best replaced by a triplet of indices: γ, d, r. Here γ denotes the irrep in question, d the *degeneracy index* if γ is two- or three-dimensional and, finally, r is a *repeat index* to cover the case in which the irrep occurs more than once in the decomposition of γ, that is, if $m_\gamma > 1$ in Eq. (12). We therefore rewrite Eq. (15) as

$$X^{(\gamma)}_{\mathrm{dr}} = S^{(\gamma)}_{\mathrm{dr,i}}X_i \qquad (3\text{–}15\text{a})$$

(Initially, the reader may feel that replacing one index by three does not lead to simplification of the problem, but, as we proceed further, it will become clear that indeed it does.)

The quantities $X^{(\gamma)}_{\mathrm{dr}}$ are called *symmetry coordinates*.* The importance of symmetry coordinates is that they transform in simple ways under the symmetry operations of the crystal. Specifically, any symmetry coordinate belonging to a one-dimensional irrep (one designated A, B or a member of an $\tilde{\mathrm{E}}$ pair in the character table) is taken into itself multiplied by a constant under all group operations. In fact, the multiplying factor is just the character given in the table (± 1 for an A or B irrep and a possible complex number for an $\tilde{\mathrm{E}}$ irrep). For a degenerate pair of symmetry coordinates belonging to a two-dimensional (E-type) irrep, the two members of the pair (called 'partners') for a given repeat index are taken into linear combinations of themselves under the group operations. To determine exactly what a given operation R does to the pair requires knowledge of the two-dimensional irrep matrices, which are easily obtained. The character tables give only the traces of these matrices, but these characters are sufficient for almost all our objectives. In a similar way, the set of three partners belonging to a given three-dimensional (T-type) irrep are taken only into linear combinations of themselves under all of the group operations.

Let us illustrate with a hypothetical example in which hypervector $\mathbf{X}$ is six-dimensional and the rep Γ generated by $\mathbf{X}$ decomposes into $2\alpha + \beta$, where α and β are both two-dimensional irreps. The block form of the

* The bar over the quantities X in Eqs. (14) and (15) was used to denote that they are symmetry coordinates. In the three-index notation, where the designation of irrep γ shows that we are dealing with symmetry coordinates, the bar is no longer necessary and is therefore dropped.

matrix $\mathbf{D}(R)$ is as follows:

$$\mathbf{D}(R) = \begin{bmatrix} \mathbf{D}^{(\alpha)}(R) & \mathbf{0} & \mathbf{0} \\ \mathbf{0} & \mathbf{D}^{(\alpha)}(R) & \mathbf{0} \\ \mathbf{0} & \mathbf{0} & \mathbf{D}^{(\beta)}(R) \end{bmatrix}$$

and each $\mathbf{0}$ is a 2×2 zero matrix. The symmetry coordinates $\bar{X}_1, \bar{X}_2, \ldots, \bar{X}_6$ can be converted to triplet indices:

$$X_{11}^{(\alpha)}, X_{21}^{(\alpha)}, X_{12}^{(\alpha)}, X_{22}^{(\alpha)}, X_1^{(\beta)}, X_2^{(\beta)}$$

where the first of the subscripts is the degeneracy index (1 or 2) and the second, the repeat index, applicable only to irrep α, which occurs twice. Thus $X_{11}^{(\alpha)}$ and $X_{21}^{(\alpha)}$ is the first pair of 'partners' that transform as irrep α and $X_{12}^{(\alpha)}$, $X_{22}^{(\alpha)}$ is the second such pair. When an index d or r is not needed, it may be omitted as in $X_1^{(\beta)}$, $X_2^{(\beta)}$ or replaced by a zero. Thus, the two coordinates of a doubly repeated *one-dimensional* irrep γ would be written $X_{01}^{(\gamma)}$, $X_{02}^{(\gamma)}$, the zero being used because there is no degeneracy index for a one-dimensional irrep, but in order to offset the second (repeat) index.

How to obtain the symmetry coordinates, that is, the transformation $\mathbf{S}$ of Eq. (14) or Eq. (15a), is one of the central problems of group theory. The formal method for so doing is called the *projection operator* method. We define an operator $P^{(\gamma)}$ by

$$P^{(\gamma)} \propto \Sigma_R \chi^{(\gamma)}(R)^* O_R \tag{3–16}$$

where $\chi^{(\gamma)}(R)^*$ is the complex conjugate of the character of operation R for the irrep γ, and O_R is an operator such that $O_R X_i$ is the coordinate which X_i becomes under the symmetry operation R. Accordingly, one must first make a listing of $O_R X_i$, that is, noting what becomes of X_i under operation R, for every X_i and every R. It can be shown that $P^{(\gamma)}$ acting on any component X_i of $\mathbf{X}$ will generate a linear combination of components that lies in the γ subspace, that is, a symmetry coordinate $X^{(\gamma)}$. (This is the reason that it is called the 'projection operator', since operating on X_i with $P^{(\gamma)}$ produces the projection of X_i into the subspace γ. If X_i lies entirely outside subspace γ, the result will be that $P^{(\gamma)} X_i = 0$.) By operating with $P^{(\gamma)}$ on all the X_i, all the symmetry coordinates belonging to irrep γ can be obtained. The proportionality in Eq. (16) is all that is necessary, since it is usually desirable to orthogonalize and normalize the symmetry coordinates such that if

$$X_{\mathrm{dr}}^{(\gamma)} = a_i X_i \tag{3–17}$$

then

$$a_i^* a_j = \delta_{ij} \tag{3–18}$$

where δ_{ij} is the Kronecker delta. This choice guarantees that **S** will be unitary. We will see, however, that, in most cases, the symmetry coordinates can be identified without the need for the projection operator method.

Let us first apply these considerations to a polar vector **X**, whose three components may be taken to be the three Cartesian coordinates x, y, z. (These coordinates transform in the same manner as the components of any polar vector.) To obtain the decomposition, we first require the character $\chi(R)$ for the three-dimensional rep Γ_{xyz} generated by these components. It is easy to show (see Problem 3–11) that the results are

$$\chi(R) = 2\cos(2\pi/n) \pm 1 \quad (\text{for } R = \mathrm{C}_n \text{ or } \mathrm{S}_n, \text{ respectively})$$

and that $\chi(\sigma) = 1$ and $\chi(\mathrm{i}) = -3$. In this way, using Eq. (12) and the character tables, we may decompose Γ_{xyz} for every point group. Except for the complex ($\tilde{\mathrm{E}}$) irreps, the symmetry coordinates are simply the coordinates x, y and z themselves; that is, the transformed vector $\bar{\mathbf{X}}$ that we seek is the same as the original vector **X** itself. It is only necessary to identify the coordinates x, y and z by the irreps to which they belong. Note that for the uniaxial groups, z always belongs to a one-dimensional irrep, while x and y are usually partners belonging to a two-dimensional E irrep. By convention, such pairs are always written within parentheses, for example as (x, y). See, for example, Table 3–2 for groups $\mathrm{C}_{4\mathrm{v}}$ and $\mathrm{D}_{3\mathrm{d}}$. This result means that with the standard choice of axes (i.e. z parallel to the major symmetry axis of the crystal), all group operations take the coordinate z into $\pm z$, while x and y are taken only into each other or into linear combinations of each other.

In the case of the complex irrep pair $\tilde{\mathrm{E}}$, the two symmetry coordinates belonging to $\tilde{\mathrm{E}}$ become $x \mp \mathrm{i}y$, as shown in Table 3–2 for the group C_6. (This result is readily checked by operating on coordinate x with the projection operator for either of the two irreps constituting $\tilde{\mathrm{E}}$.) It is worth noting here that there is no problem in the appearance of complex symmetry coordinates for the $\tilde{\mathrm{E}}$ irreps, since all physical quantities associated with these irreps will turn out to be real.

Next we wish to obtain the symmetry coordinates for components of an axial vector and for those of a second-rank symmetric tensor. To achieve this objective we must first introduce the concept of the *direct product of two reps*. Let $\mathbf{X}_1$ be the basis of a rep Γ_1 and $\mathbf{X}_2$ that of a rep Γ_2, of a given group, the reps having dimensionalities n_1 and n_2 respectively. Then consider the set of coordinates made up of all products of components, one from $\mathbf{X}_1$ and one from $\mathbf{X}_2$. It can be shown that the set of such products forms the basis of a rep Γ of dimension n_1n_2 in which the rep matrices are given by

$$\mathbf{D}(R) = \mathbf{D}_1(R) \times \mathbf{D}_2(R) \tag{3–19}$$

where $\times$ denotes a matrix direct product of the rep matrices belonging to reps Γ_1 and Γ_2. The direct product matrix $\mathbf{D}(R)$ contains all products of an element of $\mathbf{D}_1(R)$ by an element of $\mathbf{D}_2(R)$ arranged in a suitable 'dictionary' manner. (Details of this type of matrix multiplication are given in the standard references.) The important result for present purposes, however, is that the value of the character of the matrix $\mathbf{D}(R)$ is given by

$$\chi(R) = \chi_1(R)\chi_2(R) \tag{3–20}$$

that is, it is obtained by ordinary numerical multiplication of the characters of $\mathbf{D}_1(R)$ and $\mathbf{D}_2(R)$. Thus, if the characters are already known for Γ_1 and Γ_2, the characters of Γ are easily obtained from Eq. (20).

Let us apply these concepts to a basis which is the set of products of the components of two different polar vectors $\mathbf{v}$ and $\mathbf{w}$, each with three components. Clearly, this basis has nine components which, in fact, transform like the components of a second-rank tensor. It is convenient to form linear combinations which are the symmetric and antisymmetric products, as follows:

$$\left.\begin{aligned} \text{Symmetric:}\, v_iw_j + v_jw_i \text{ (six components)} \\ \text{Antisymmetric:}\, v_iw_j - v_jw_i \text{ (three components)} \end{aligned}\right\} \tag{3–21}$$

(The symmetric products allow $i = j$ while the antisymmetric do not.) The antisymmetric products are, in fact, the three components of the vector cross product $\mathbf{R} = \mathbf{v} \times \mathbf{w}$ constituting an *axial vector**

$$\left.\begin{aligned} R_x &= v_2w_3 - v_3w_2 \\ R_y &= v_3w_1 - v_1w_3 \\ R_z &= v_1w_2 - v_2w_1 \end{aligned}\right\} \tag{3–22}$$

* The reader should not confuse the symbol for a general axial vector, $\mathbf{R} = (R_x, R_y, R_z)$ with the symbol R used for a representative group operation. Unfortunately, both have become standard notation.

The symmetric products are the six components of a second-rank symmetric tensor. The symmetric products give the same results whether **v** and **w** are the same or different. Therefore, it is convenient to use the products of the vector x, y, z with itself to form the six components x^2, y^2, z^2, yz, zx, xy. (Of course, the antisymmetric products vanish if **v** and **w** are the same.)

The importance of making the separation into symmetric and antisymmetric products is that under any unitary transformation symmetric components only transform into linear combinations of symmetric components, and similarly for antisymmetric components. (See Problem 3–13.) Such a separation therefore takes us part way toward reducing the rep Γ, generated by the products $v_i w_j$, by reducing the rep into a two-block form, that is, the rep Γ has been partially decomposed into

$$\Gamma = \Gamma_{aS} + \Gamma_S \tag{3–23}$$

where Γ_{aS} is the rep formed by the three components of an axial vector (or antisymmetric second-rank tensor) and Γ_S is the rep formed by the six components of a second-rank symmetric tensor. To complete the reduction we need the appropriate characters. From Eq. (20) it is clear that for rep Γ the characters are $[\chi_{xyz}(R)]^2$ where $\chi_{xyz}(R)$ are the characters of the rep Γ_{xyz}. The characters of the antisymmetric and symmetric reps, respectively, are shown in standard references (see e.g. Burns, 1977, Section 6.1) to be

$$\left.\begin{aligned} \chi_{aS}(R) &= 1/2\{[\chi_{xyz}(R)]^2 - \chi_{xyz}(R^2)\} \\ \chi_S(R) &= 1/2\{[\chi_{xyz}(R)]^2 + \chi_{xyz}(R^2)\} \end{aligned}\right\} \tag{3–24}$$

where $\chi_{xyz}(R^2)$ means the character for the operation R^2 (for example if $R = C_4$, $R^2 = C_2$, or if $R = \sigma$, $R^2 = E$). Of course, $\chi_{aS}(R)$ and $\chi_S(R)$ add up to $[\chi_{xyz}(R)]^2$, the character of rep Γ, as they must, based on Eq. (23). By knowing the characters of the reps Γ_{aS} and Γ_S generated by the antisymmetric and symmetric products we may decompose each of these reps into irreps by using Eq. (12) together with the character table of the irreps.

Although it is not obvious from Eq. (24), it turns out (see Problem 3–14) that for any pure rotation, that is, $R = C_n$,

$$\chi_{aS}(R) = \chi_{xyz}(R) \tag{3–25}$$

while for all symmetry operations involving inversion or mirror planes, that is, $R = i, \sigma, S_n$,

$$\chi_{aS}(R) = -\chi_{xyz}(R) \tag{3–26}$$

This result illustrates the validity of thinking of a polar vector as an arrow and of an axial vector as a line of definite orientation combined with a sense of rotation, as shown in Fig. A–1. Under a pure rotation, in which the handedness of the axes is unchanged, these two types of vector transform in the same way. Under any operation (reflection, inversion or S_n) in which the handedness of the axes is reversed, the components of an axial vector transform oppositely to those of a polar vector. The three symmetry coordinates of an axial vector are usually presented to the right of the character tables as the quantities R_x, R_y and R_z, of Eq. (22), as shown in Table 3–2 for groups C_6, C_{4v} and D_{3d}. Again, these components are themselves symmetry coordinates in all cases except for the $\tilde{E}$ irreps. In accordance with the above discussion, we note that, for any group involving rotations only (as e.g. C_6), R_z transforms as z, R_x as x and R_y as y. When operations i, σ and/or S_n are present, however, the axial and polar vector components transform differently. Specifically in a group which possesses inversion symmetry, as, for example, D_{3d}, the components of a polar vector transform as irreps of odd parity (subscript u) while those of an axial vector transform as irreps of even parity (subscript g). This is because a polar vector is reversed in the inversion operation while an axial vector is not (as can readily be seen from the diagrams of Fig. A–1). However, except for this difference in parity, the irreps correspond (e.g. for group D_{3d}: A_{2g} and E_g for Γ_{xyz} versus A_{2u} and E_u for Γ_{aS}) because the two types of vector transform in the same way under the pure rotations, C_n.

Turning now to the rep Γ_S generated by the six components of a symmetric second-rank tensor, we may make use of the second of Eqs. (24) to obtain the characters $\chi_s(R)$ and then carry out the decomposition through Eq. (12). The projection operator method, Eq. (16), may then be used to obtain the symmetry coordinates, as illustrated, for example, by Problem 3–15.

The task can often be greatly simplified by considering, separately, the direct products of components of the vector x, y, z that belong to the various irreps, and utilizing the character tables. We illustrate for the uniaxial group D_{3d}. Here the component z belongs to the one-dimensional irrep A_{2u}. Therefore, z^2 belongs to $A_{2u} \times A_{2u}$. The characters of this direct product rep are obtainable from Eq. (20) and are clearly given by $\chi(R) = 1$, for all R, that is, the product rep is the totally symmetric irrep A_{1g}. Thus z^2 is invariant. But for *any* group, $x^2 + y^2 + z^2$ is invariant. Thus $x^2 + y^2$ must also be invariant. Now, the products xz and yz are obtainable from the direct product $A_{2u} \times E_u$, since x and y belong to E_u.

From the products of the characters, it is clear that $A_{2u} \times E_u = E_g$. Thus xz and yz as a pair form the basis of the irrep E_g. The remaining components involve products of x and y with each other. The rep generated by such products is $E_u \times E_u$, which, by taking the products of characters and carrying out the decomposition gives

$$E_u \times E_u = A_{1g} + A_{2g} + E_g$$

Now, among the products of x, y with itself there are three symmetric and one antisymmetric product. The antisymmetric one corresponds to R_z and clearly belongs to A_{2g}. Of the symmetric products, we have already decided that $x^2 + y^2$ belongs to A_{1g}. The remaining two product symmetry coordinates must be $x^2 - y^2$ and xy and these, together, are the basis of the irrep E_g. Thus all six symmetry coordinates have been identified without the need to employ the projection operator method.

It is usual to place these symmetry coordinates of Γ_S along with those of Γ_{xyz} and Γ_{aS} to the right of the character table opposite the appropriate irreps, in the manner shown in Table 3–2 for point groups C_6, C_{4v} and D_{3d}. The case of D_{3d} illustrates a general rule for all groups that possess inversion symmetry, namely, the components of a second-rank symmetric tensor always belong to irreps of even parity (those labeled sub g). The reason follows from the fact that, as already mentioned, vector components (x, y, z) belong to u-type irreps (odd parity). Further, from the character product rule, Eq. (20), it follows that the direct product of two irreps of odd parity must be even (simply because $(-1)\cdot(-1) = +1$). In abbreviated notation this may be stated as

$$u \times u = g \tag{3–27}$$

Thus for groups that have inversion symmetry, polar vector and second-rank symmetric tensor components can never belong to the same irrep. (We will see later, Section 7–1–1, that this statement tells us immediately that such crystals cannot be piezoelectric.)

3–5 Concept of similarity of orientation

The determination of the symmetry coordinates is not always unambiguous. In the case of two- (or three-) dimensional irreps of type E (or T) there are, in fact, an infinite number of choices for the symmetry coordinates corresponding to that irrep. For example, if x and y belong to the irrep E, equally suitable are any pair of coordinates obtained by an orthogonal transformation

$$\mathbf{X}' = \mathbf{S}\mathbf{X} \tag{3–28}$$

where $\mathbf{X}$ represents the vector x, y, $\mathbf{X}'$ the new vector (both expressed as column vectors) and $\mathbf{S}$ is a 2×2 orthogonal matrix. Such new coordinates will be orthogonal to each other as well as to any other symmetry coordinates belonging to different irreps, in this case the coordinate z. What the transformation (28) implies is a rotation, about the z-axis in this case, to a new orthogonal set of coordinates in the $x-y$-plane, that is, to a new orientation of the x- and y-axes.

If only one pair of symmetry coordinates of interest in a given problem belongs to irrep E, it does not matter what orientation we select, and so one customarily chooses the simplest, for example x and y alone, as against some linear combination of x and y given by Eq. (28). As soon as there is a second pair of coordinates, relevant to the problem at hand, that belongs to the same degenerate irrep, it becomes advantageous to make sure that the two pairs are *similarly oriented*, that is, that they transform in a corresponding fashion under all symmetry operations. The reason for this specification is that we require each group operation to produce exactly the same transformation of each pair, as we will discuss in Chapter 4.

What we mean by similar orientation is best illustrated by example. Consider the character table for the group C_{4v}, presented in Table 3–2. We see that the pairs (x, y), (R_x, R_y) and (xz, yz) all belong to the two-dimensional irrep E. Let us fix the orientation with the pair (x, y). We then ask: do the pairs (R_x, R_y) and (xz, yz) have the same orientation as (x, y)? In the case of xz and yz the answer is simple: z is invariant (i.e. it belongs to irrep A_1) so that xz must transform exactly as x and yz as y. In the case of R_x, R_y we may immediately suspect that the orientations are not the same, since axial vectors transform differently from polar vectors under symmetry operations that are not pure rotations, and the present group includes the vertical (i.e. containing the C_4 axis) mirror planes σ_v and σ_d. In order to determine how the relevant quantities transform, we need only consider the generating elements of the group: C_4 and σ_v. Under C_4 (a rotation of 90° about the z-axis):

$$x \rightarrow y, \quad y \rightarrow -x, \quad z \rightarrow z$$

Since this result applies to any polar vector components, we may apply it to R_x and R_y through the first two of Eqs. (22) to obtain

$$R_x \rightarrow R_y \text{ and } R_y \rightarrow -R_x$$

Table 3–2. *The character tables of the 32 crystallographic point groups and of two infinite groups*

The non-axial groups

C_1	E
A	1

C_i	E	σ_h
A_g	1	1
A_u	1	−1

C_s	E	σ
A′	1	1
A″	1	−1

The C_n groups

C_2	E	C_2
A	1	1
B	1	−1

C_3	E	C_3	C_3^2
A	1	1	1
$\tilde{E}$	1	ε	ε^*
	1	ε^*	ε

$\varepsilon = \exp(2\pi i/3)$

C_4	E	C_4	C_2	C_4^3
A	1	1	1	1
B	1	−1	1	−1
$\tilde{E}$	1	i	−1	−i
	1	−i	−1	i

C_6	E	C_6	C_3	C_2	C_3^2	C_6^5		
A	1	1	1	1	1	1	z; R_z	x^2+y^2; z^2
B	1	−1	1	−1	1	−1	–	–
$\tilde{E}_1$	1	ε	$-\varepsilon^*$	−1	$-\varepsilon$	ε^*	$x-iy$; R_x-iR_y	$xz-iyz$
	1	ε^*	$-\varepsilon$	−1	$-\varepsilon^*$	ε	$x+iy$; R_x+iR_y	$xz+iyz$
$\tilde{E}_2$	1	$-\varepsilon^*$	$-\varepsilon$	1	$-\varepsilon^*$	$-\varepsilon$	–	$(x^2-y^2)-2ixy$
	1	$-\varepsilon$	$-\varepsilon^*$	1	$-\varepsilon$	$-\varepsilon^*$	–	$(x^2-y^2)+2ixy$

$\varepsilon = \exp(2\pi i/6)$

The C_{nv} groups

C_{2v}	E	C_2	σ_v^{xz}	σ_v^{yz}
A_1	1	1	1	1
A_2	1	1	−1	−1
B_1	1	−1	1	−1
B_2	1	−1	−1	1

C_{3v}	E	$2C_3$	$3\sigma_v$
A_1	1	1	1
A_2	1	1	−1
E	2	−1	0

Table 3–2. *(cont.)*

C_{4v}	E	$2C_4$	C_2	$2\sigma_v$	$2\sigma_d$		
A_1	1	1	1	1	1	z	$x^2+y^2; z^2$
A_2	1	1	1	−1	−1	R_z	–
B_1	1	−1	1	1	−1	–	x^2-y^2
B_2	1	−1	1	−1	1	–	xy
E	2	0	−2	0	0	$(x, y); (R_x, R_y)$	(xz, yz)

C_{6v}	E	$2C_6$	$2C_3$	C_2	$3\sigma_v$	$3\sigma_d$
A_1	1	1	1	1	1	1
A_2	1	1	1	1	−1	−1
B_1	1	−1	1	−1	1	−1
B_2	1	−1	1	−1	−1	1
E_1	2	1	−1	−2	0	0
E_2	2	−1	−1	2	0	0

The C_{nh} groups

C_{2h}	E	C_2	i	σ_h
A_g	1	1	1	1
B_g	1	−1	1	−1
A_u	1	1	−1	−1
B_u	1	−1	−1	1

C_{3h}	E	C_3	C_3^2	σ_h	S_3	S_3^5
A′	1	1	1	1	1	1
$\tilde{E}'$	1	ε	ε^*	1	ε	ε^*
	1	ε^*	ε	1	ε^*	ε
A″	1	1	1	−1	−1	−1
$\tilde{E}''$	1	ε	ε^*	−1	$-\varepsilon$	$-\varepsilon^*$
	1	ε^*	ε	−1	$-\varepsilon^*$	$-\varepsilon$

$\varepsilon = \exp(2\pi i/3)$

C_{4h}	E	C_4	C_2	C_4^3	i	S_4^3	σ_h	S_4
A_g	1	1	1	1	1	1	1	1
B_g	1	−1	1	−1	1	−1	1	−1
$\tilde{E}_g$	1	i	−1	−i	1	i	−1	−i
	1	−i	−1	i	1	−i	−1	i
A_u	1	1	1	1	−1	−1	−1	−1
B_u	1	−1	1	−1	−1	1	−1	1
$\tilde{E}_u$	1	i	−1	−i	−1	−i	1	i
	1	−i	−1	i	−1	i	1	−i

Table 3–2. *(cont.)*

C_{6h}	E	C_6	C_3	C_2	C_3^2	C_6^5	i	S_3^5	S_6^5	σ_h	S_6	S_3
A_g	1	1	1	1	1	1	1	1	1	1	1	1
B_g	1	−1	1	−1	1	−1	1	−1	1	−1	1	−1
$\tilde{E}_{1g}$	1	ε	$-\varepsilon^*$	−1	$-\varepsilon$	ε^*	1	ε	$-\varepsilon^*$	−1	$-\varepsilon$	ε^*
	1	ε^*	$-\varepsilon$	−1	$-\varepsilon^*$	ε	1	ε^*	$-\varepsilon$	−1	$-\varepsilon^*$	ε
$\tilde{E}_{2g}$	1	$-\varepsilon^*$	$-\varepsilon$	1	$-\varepsilon^*$	$-\varepsilon$	1	$-\varepsilon^*$	$-\varepsilon$	1	$-\varepsilon^*$	$-\varepsilon$
	1	$-\varepsilon$	$-\varepsilon^*$	1	$-\varepsilon$	$-\varepsilon^*$	1	$-\varepsilon$	$-\varepsilon^*$	1	$-\varepsilon$	$-\varepsilon^*$
A_u	1	1	1	1	1	1	−1	−1	−1	−1	−1	−1
B_u	1	−1	1	−1	1	1	−1	1	−1	1	−1	1
$\tilde{E}_{1u}$	1	ε	$-\varepsilon^*$	−1	$-\varepsilon$	ε^*	−1	$-\varepsilon$	ε^*	1	ε	$-\varepsilon^*$
	1	ε^*	$-\varepsilon$	−1	$-\varepsilon^*$	ε	−1	$-\varepsilon^*$	ε	1	ε^*	$-\varepsilon$
$\tilde{E}_{2u}$	1	$-\varepsilon^*$	$-\varepsilon$	1	$-\varepsilon^*$	$-\varepsilon$	−1	ε^*	ε	−1	ε^*	ε
	1	$-\varepsilon$	$-\varepsilon^*$	1	$-\varepsilon$	$-\varepsilon^*$	−1	ε	ε^*	−1	ε	ε^*

$\varepsilon = \exp(2\pi i/6)$

The D_n *groups*

D_2	E	$C_2(z)$	$C_2(y)$	$C_2(x)$
A	1	1	1	1
B_1	1	1	−1	−1
B_2	1	−1	1	−1
B_3	1	−1	−1	1

D_3	E	$2C_3$	$3C_2$
A_1	1	1	1
A_2	1	1	−1
E	2	−1	0

D_4	E	$2C_4$	$C_2(=C_4^2)$	$2C_2'$	$2C_2''$
A_1	1	1	1	1	1
A_2	1	1	1	−1	−1
B_1	1	−1	1	1	−1
B_2	1	−1	1	−1	1
E	2	0	−2	0	0

D_6	E	$2C_6$	$2C_3$	C_2	$3C_2'$	$3C_2''$
A_1	1	1	1	1	1	1
A_2	1	1	1	1	−1	−1
B_1	1	−1	1	−1	1	−1
B_2	1	−1	1	−1	−1	1
E_1	2	1	−1	−2	0	0
E_2	2	−1	−1	2	0	0

Table 3–2. *(cont.)*

The D_{nh} *groups*

D_{2h}	E	$C_2(z)$	$C_2(y)$	$C_2(x)$	i	σ^{xy}	σ^{xz}	σ^{yz}
A_g	1	1	1	1	1	1	1	1
B_{1g}	1	1	−1	−1	1	1	−1	−1
B_{2g}	1	−1	1	−1	1	−1	1	−1
B_{3g}	1	−1	−1	1	1	−1	−1	1
A_u	1	1	1	1	−1	−1	−1	−1
B_{1u}	1	1	−1	−1	−1	−1	1	1
B_{2u}	1	−1	1	−1	−1	1	−1	1
B_{3u}	1	−1	−1	1	−1	1	1	−1

D_{3h}	E	$2C_3$	$3C_2$	σ_h	$2S_3$	$3\sigma_v$
A_1'	1	1	1	1	1	1
A_2'	1	1	−1	1	1	−1
E'	2	−1	0	2	−1	0
A_1''	1	1	1	−1	−1	−1
A_2''	1	1	−1	−1	−1	1
E''	2	−1	0	−2	1	0

D_{4h}	E	$2C_4$	C_2	$2C_2'$	$2C_2''$	i	$2S_4$	σ_h	$2\sigma_v$	$2\sigma_d$
A_{1g}	1	1	1	1	1	1	1	1	1	1
A_{2g}	1	1	1	−1	−1	1	1	1	−1	−1
B_{1g}	1	−1	1	1	−1	1	−1	1	1	−1
B_{2g}	1	−1	1	−1	1	1	−1	1	−1	1
E_g	2	0	−2	0	0	2	0	−2	0	0
A_{1u}	1	1	1	1	1	−1	−1	−1	−1	−1
A_{2u}	1	1	1	−1	−1	−1	−1	−1	1	1
B_{1u}	1	−1	1	1	−1	−1	1	−1	−1	1
B_{2u}	1	−1	1	−1	1	−1	1	−1	1	−1
E_u	2	0	−2	0	0	−2	0	2	0	0

D_{6h}	E	$2C_6$	$2C_3$	C_2	$3C_2'$	$3C_2''$	i	$2S_3$	$2S_6$	σ_h	$3\sigma_d$	$3\sigma_v$
A_{1g}	1	1	1	1	1	1	1	1	1	1	1	1
A_{2g}	1	1	1	1	−1	−1	1	1	1	1	−1	−1
B_{1g}	1	−1	1	−1	1	−1	1	−1	1	−1	1	−1
B_{2g}	1	−1	1	−1	−1	1	1	−1	1	−1	−1	1
E_{1g}	2	1	−1	−2	0	0	2	1	−1	−2	0	0
E_{2g}	2	−1	−1	2	0	0	2	−1	−1	2	0	0
A_{1u}	1	1	1	1	1	1	−1	−1	−1	−1	−1	−1
A_{2u}	1	1	1	1	−1	−1	−1	−1	−1	−1	1	1
B_{1u}	1	−1	1	−1	1	−1	−1	1	−1	1	−1	1
B_{2u}	1	−1	1	−1	−1	1	−1	1	−1	1	1	−1
E_{1u}	2	1	−1	−2	0	0	−2	−1	1	2	0	0
E_{2u}	2	−1	−1	2	0	0	−2	1	1	−2	0	0

Table 3–2. *(cont.)*

The D_{nd} *groups*

D_{2d}	E	$2S_4$	C_2	$2C_2'$	$2\sigma_d$
A_1	1	1	1	1	1
A_2	1	1	1	−1	−1
B_1	1	−1	1	1	−1
B_2	1	−1	1	−1	1
E	2	0	−2	0	0

D_{3d}	E	$2C_3$	$3C_2$	i	$2S_6$	$3\sigma_d$		
A_{1g}	1	1	1	1	1	1	–	$x^2+y^2; z^2$
A_{2g}	1	1	−1	1	1	−1	R_z	–
E_g	2	−1	0	2	−1	0	(R_x, R_y)	$(x^2-y^2, xy); (xz, yz)$
A_{1u}	1	1	1	−1	−1	−1	–	
A_{2u}	1	1	−1	−1	−1	1	z	
E_u	2	−1	0	−2	1	0	(x, y)	

The S_n *groups*

S_4	E	S_4	C_2	C_4^3
A	1	1	1	1
B	1	−1	1	−1
$\tilde{E}$	1	i	−1	−i
	1	−i	−1	i

S_6	E	C_3	C_3^2	i	S_6^5	S_6
A_g	1	1	1	1	1	1
$\tilde{E}_g$	1	ε	ε^*	1	ε	ε^*
	1	ε^*	ε	1	ε^*	ε
A_u	1	1	1	−1	−1	−1
$\tilde{E}_u$	1	ε	ε^*	−1	$-\varepsilon$	$-\varepsilon^*$
	1	ε^*	ε	−1	$-\varepsilon^*$	$-\varepsilon$

$\varepsilon = \exp(2\pi i/3)$

The cubic groups

T	E	$4C_3$	$4C_3^2$	$3C_2$
A	1	1	1	1
$\tilde{E}$	1	ε	ε^*	1
	1	ε^*	ε	1
T	3	0	0	−1

$\varepsilon = \exp(2\pi i/3)$

O	E	$8C_3$	$6C_2'$	$6C_4$	$3C_2\ (\equiv C_4^2)$
A_1	1	1	1	1	1
A_2	1	1	−1	−1	1
E	2	−1	0	0	2
T_1	3	0	−1	1	−1
T_2	3	0	1	−1	−1

Table 3–2. *(cont.)*

O_h	E	$8C_3$	$6C_2'$	$6C_4$	$3C_2\ (= C_4^2)$	i	$6S_4$	$8S_6$	$3\sigma_h$	$6\sigma_d$
A_{1g}	1	1	1	1	1	1	1	1	1	1
A_{2g}	1	1	−1	−1	1	1	−1	1	1	−1
E_g	2	−1	0	0	2	2	0	−1	2	0
T_{1g}	3	0	−1	1	−1	3	1	0	−1	−1
T_{2g}	3	0	1	−1	−1	3	−1	0	−1	1
A_{1u}	1	1	1	1	1	−1	−1	−1	−1	−1
A_{2u}	1	1	−1	−1	1	−1	1	−1	−1	1
E_u	2	−1	0	0	2	−2	0	1	−2	0
T_{1u}	3	0	−1	1	−1	−3	−1	0	1	1
T_{2u}	3	0	1	−1	−1	−3	1	0	1	−1

T_h	E	$4C_3$	$4C_3^2$	$3C_2$	i	$4S_6^5$	$4S_6$	$3\sigma_h$
A_g	1	1	1	1	1	1	1	1
$\tilde{E}_g$	1	ε	ε^*	1	1	ε	ε^*	1
	1	ε^*	ε	1	1	ε^*	ε	1
T_g	3	0	0	−1	3	0	0	−1
A_u	1	1	1	1	−1	−1	−1	−1
$\tilde{E}_u$	1	ε	ε^*	1	−1	$-\varepsilon$	$-\varepsilon^*$	−1
	1	ε^*	ε	1	−1	$-\varepsilon^*$	$-\varepsilon$	−1
T_u	3	0	0	−1	−3	0	0	1

T_d	E	$8C_3$	$3C_2$	$6S_4$	$6\sigma_d$
A_1	1	1	1	1	1
A_2	1	1	1	−1	−1
E	2	−1	2	0	0
T_1	3	0	−1	1	−1
T_2	3	0	−1	−1	1

Two infinite groups

$C_{\infty v}$	E	$2C(\phi)$	$\infty\sigma_v$
A_1	1	1	1
A_2	1	1	−1
E_1	2	$2\cos\phi$	0
E_2	2	$2\cos 2\phi$	0
⋮	⋮	⋮	⋮
E_n	2	$2\cos n\phi$	0
⋮	⋮	⋮	⋮

$R(3)$	E	$\infty C(\theta)$
A	1	1
T	3	$\sin(3\theta/2)/\sin(\theta/2)$
H	5	$\sin(5\theta/2)/\sin(\theta/2)$
⋮	⋮	⋮

This result still allows for two possibilities: either R_x corresponds to x and R_y to y, or R_y corresponds to x and $-R_x$ to y.* In order to choose between these two possibilities consider the vertical mirror plane σ_v^{yz}, that is, the yz-plane. Under this symmetry operation we find

$$x \rightarrow -x, \quad y \rightarrow y, \quad z \rightarrow z$$

When this is applied to the components of v_i and w_j of Eqs. (22) we find

$$R_x \rightarrow R_x, \quad R_y \rightarrow -R_y$$

This result is only consistent with the second of the above alternatives, namely, $R_y \sim x$, $-R_x \sim y$. Thus, the correct way to write this pair is $(R_y, -R_x)$ if an orientation similar to (x, y) is desired. The fact that the correct answer is not (R_x, R_y) could have been anticipated because the group C_{4v} contains operations (the mirror planes) that change the handedness of the axes; therefore x, y, z and R_x, R_y, R_z cannot transform in the same way with respect to all the group operations.

The character tables given in most textbooks (of which the cases of C_6, C_{4v} and D_{3d} in Table 3–2 are examples) do not specify pairs belonging to irreps E (or triplets in the case of irreps T of the cubic groups) with similarity of orientation in mind. For our purposes, however, this matter is of such importance that separate tables are to be presented (see Section 4–3 and Appendix E) in which similarity of orientation is fully taken into account.

With the completion of this review of group (representaton) theory, we are now prepared to undertake the analysis of matter tensors, taking account of crystal symmetry.

Problems

This problem set is intended primarily for the reader whose familiarity with group theory is minimal.

3–1. Consider the multiplication table for the non-Abelian abstract group of order 6 (page 33). Show that there are three classes: E by itself; A and B; C, D and F. Also show that A and C can be taken as generators. What other pair of generators could have been selected?

* A change in both signs, that is, $-R_y \sim x$ and $R_x \sim y$ is not considered a different orientation, any more than changing from x, y to $-x$, $-y$. The reason is that a negative pair will transform in exactly the same way under all symmetry operations as the corresponding positive pair. Thus, in considering the possible orientations of the pair R_x and R_y to match the orientation (x, y), one finds that there are, *a priori*, just four possibilities: (R_x, R_y), $(R_x, -R_y)$, (R_y, R_x) and $(R_y, -R_x)$.

3–2. In the diagrams of Fig. 3–2 for group D_{2d}, label each of the six dots and circles with the appropriate group operation taken from Table 3–1; start from the point in the first quadrant and above the x–y-plane as the one corresponding to the identity operation, E.

3–3. Do the same as in Problem 3–2 for groups C_{3v} and C_{4h}.

3–4. Show that the S_6 group is a cyclic group obtained by repeated application of the S_6 operation.

3–5. Show that groups D_3 and C_{3v} obey the multiplication table of the abstract group on page 33. Shown which symmetry operations correspond to each of the abstract symbols A, . . ., F in each case.

3–6. Show that in an Abelian group, every operation is in a class by itself.

3–7. Show that a change of basis vectors through Eq. (6) leads to the similarity transformaton of rep matrices according to Eq. (7).

3–8. Using the fact that $\chi(\mathbf{AB}) = \chi(\mathbf{BA})$ (where **A** and **B** are two matrices of the same dimensionality), show that the character of a matrix does not change under a similarity transformation.

3–9. Use the orthogonality theorem for characters, Eq. (13), together with Eq. (11), to prove the decomposition theorem, Eq. (12).

3–10. Generate the character table for group C_{4v} utilizing theorems (1)–(5), pages 43–4, especially Eqs. (10) and (13).

3–11. Starting from the full $\mathbf{D}(R)$ matrices for vector x, y, z, show that the characters of the three-dimensional rep Γ_{xyz} are $\chi(R) = 2\cos(2\pi/n) \pm 1$ for $R = C_n$ or S_n, respectively, that $\chi(\sigma) = 1$, and $\chi(i) = -3$. Note that $\chi(\sigma)$ is the same for any mirror plane (Why?), so that you are free to choose the most convenient one for your calculation.

3–12. Using the projection operator on coordinate x, show that the symmetry coordinates belonging to $\tilde{E}_1$ in group C_6 are $x \mp iy$. (Begin by making a listing of $O_R x$ for every operation R.) Also show that $P^{(A)}x = P^{(B)}x = P^{(\tilde{E}_2)}x = 0$, meaning that there are no symmetry coordinates involving x that belong to irreps A, B or $\tilde{E}_2$.

3–13. Show that under a unitary transformation the components of the symmetric products of two vectors only transform into symmetric components, and similarly for antisymmetric components.

3–14. Using the results of Problem 3–11, and with the aid of Eqs. (24), verify Eqs. (25) and (26).

3–15. Use the projection operator method to obtain the symmetry coordinates of the symmetric products x^2, y^2, ..., xy in the group C_{4v}, and identify the irreps to which they belong. Then repeat the problem using the simple method of taking products of appropriate irreps as described on page 53. (Include working out the decomposition of $E \times E$.)

3–16. Do the same as in Problem 3–15 for the group C_6.

3–17. Noting the definition of a pseudoscalar (see Eq. (A–39)), determine, by inspection, to which irrep a pseudoscalar quantity belongs for each point group. (Note that it belongs to the totally symmetric irrep only for point groups C_n and D_n. Why?)

3–18. In the group D_{3d} the pairs (R_x, R_y), (yz, xz) and $(x^2 - y^2, xy)$ all belong to the irrep E. Taking the pair (yz, xz) to set the orientation, obtain the other two pairs in a similar orientation. (Hint: use the two generating operations: C_3 and $C_2(x)$.)

4

Linear relations treated group theoretically

4–1 Introduction and Neumann's principle

In the application of group theory to crystal properties, we will be concerned with a relation of the type

$$\mathbf{Y} = \mathbf{KX} \tag{4–1}$$

where **Y** and **X** are vector or tensor physical quantities (e.g. a 'response' and 'force', respectively) and **K** is the corresponding matter tensor. The property **K** can represent a principal effect if **X** and **Y** are conjugate quantities, or it can be a cross effect when **X** and **Y** are not conjugate (see Chapters 1 and 2).

The various matter tensors that we have considered show certain symmetries (invariances under interchange of indices) which stem from

(a) the nature of the physical quantities involved;
(b) thermodynamic reciprocity relations (for the equilibrium properties);
(c) Onsager reciprocity relations (for transport properties).

These have been called 'intrinsic symmetries' and are listed in Tables 1–1 and 2–1, which summarize the various tensors considered. The maximum number of independent coefficients, also listed in these tables, is arrived at from a consideration of the tensor character of the property in question together with its intrinsic symmetries. This maximum number is the number of independent constants for the case of a triclinic crystal belonging to group C_1, that is, without any crystal symmetry. In the usual cases of crystals which possess some symmetry, the matter tensor in question will generally be further simplified in accordance with *Neumann's principle* which states:

Any symmetry exhibited by the point group of the crystal (the macroscopic symmetry: rotations, reflections, and inversion) must also be possessed by every physical property of the crystal.

P. Curie preferred to state this principle in terms of asymmetry rather than symmetry. His statement was: 'No asymmetry can be manifested in a physical property of a crystal that is not already present in the crystal itself.'

Note that there is nothing in either statement to prohibit a physical property from possessing *more* symmetry than does the crystal which exhibits that property. For example, properties that are centrosymmetric in nature can be observed for a crystal that does not possess a center of symmetry. In general, the principle implies that the symmetry of the crystal either matches that of the property or is a subgroup of the symmetry exhibited by the property.

Neumann's principle is an intuitive statement which, with a little thought, becomes obvious. The application of this principle, however, has a profound effect on the study of anisotropic properties of crystals. We will see that the effect of crystal symmetry is to decrease the number of constants in a given matter tensor to below the maximum number given in Tables 1–1 and 2–1, by requiring that some coefficients be equal to zero and that others be equal to each other or to combinations of others.

It is important to emphasize that symmetry alone cannot tell us whether a given coefficient, say K_{ij}, is large or small; such information can only come from a knowledge of the detailed physics of the given material for the property under consideration. But it can be stated that a coefficient can only be strictly zero if that is required by symmetry; otherwise, the coefficient may be small, and even experimentally undetectable, but it will nevertheless not be zero.

The major objective of the remainder of this book is to examine the consequences of Neumann's principle for all possible crystal symmetries by employing the powerful mathematical methods of group theory.

4–2 Tensor quantities as hypervectors

For a group theoretical treatment, it is desirable to convert **Y** and **X** of Eq. (1) to hypervector form. For a polar vector, this involves no change, with the components identified by the subscripts $i = 1, 2, 3$. Similarly, for an axial vector, it has been shown (see Eq. (3–22) and Appendix A) that single-index notation, $i = 1, 2, 3$, may be used in place of the two-index notation for an antisymmetric second-rank tensor. Now consider a symmetric second-rank tensor α_{ij}. It has six independent components, α_{11}, α_{22}, α_{33}, α_{23}, α_{31}, α_{12}, which transform the same way as the products x^2, y^2, z^2, yz, zx, xy. We have already seen that these six components can serve as the basis for a six-dimensional rep. Although, in the preceding chapter, we have shown how to decompose such a rep into irreps and how to obtain the symmetry coordinates, we have not dealt with the question of creating a proper hypervector out of these components. We do this now in a manner that is suitable for generalization to other tensors.

Consider a set of basis functions ψ_i $(i = 1, 2, \ldots, n)$ from which we wish to construct new basis functions as products

$$\phi_k = \psi_i \psi_j$$

It is easily shown (since $\psi_i \psi_j = \psi_j \psi_i$ must be counted only once) that the single index k runs from 1 to $(n^2 + n)/2$. To provide a suitable basis, a hypervector must have an absolute magnitude that is invariant under any unitary transformation. Thus, for example, the set ψ_i must obey

$$\psi_i^* \psi_i = \text{constant} \tag{4–2}$$

with respect to such a transformation. If the ϕ_k are defined as products in the above manner, however, it is easy to show from Eq. (2) that $\phi_k^* \phi_k \neq \text{constant}$. The correct way to form the products, which was pointed out by Weyl (1950) and will be referred to as the *Weyl normalization*, is as follows:

$$\phi_k = \begin{cases} \psi_i \psi_j & (i = j) \\ \sqrt{2}\psi_i \psi_j & (i \neq j) \end{cases} \tag{4–3}$$

With this definition, it is clear that $\phi_k^* \phi_k = [\psi_i^* \psi_i]^2 = \text{constant}$, that is, that ϕ_k defined in this way provides a suitable set of basis functions for a unitary rep.

In the case where $\psi_i = x, y, z$, the definition (3) gives the products x^2, y^2, z^2, $\sqrt{2}yz$, $\sqrt{2}zx$, $\sqrt{2}xy$ as the components of a six-vector. Correspondingly, the components of any symmetric second-rank tensor α_{ij} can be converted to six-vector single-index notation α_k as follows:

Tensor:	α_{11}	α_{22}	α_{33}	$\sqrt{2}\alpha_{23}$	$\sqrt{2}\alpha_{31}$	$\sqrt{2}\alpha_{12}$	(4–4)
Six-vector:	α_1	α_2	α_3	α_4	α_5	α_6	

The reader should note that this conversion differs from that which gives the one-index engineering stresses and strains (see page 8).

The argument that led to Eq. (3) is readily generalized to the case of triple products as follows:

$$\phi_l = \begin{cases} \psi_i \psi_j \psi_k & \text{if } i = j = k \\ \sqrt{3}\psi_i \psi_j \psi_k & \text{if two indices are equal} \\ \sqrt{6}\psi_i \psi_j \psi_k & \text{if } i \neq j \neq k \end{cases} \tag{4–5}$$

where the single index l runs from 1 to $(n^3 + 3n^2 + 2n)/6$. The most important application is to the case of triple products of polar vector components, where $l = 1, \ldots, 10$ (since $n = 3$). In this case, a ten-vector: $\beta_1, \ldots, \beta_{10}$ is defined with components given by the products: x^3, y^3, z^3, $\sqrt{3}x^2y$, $\sqrt{3}x^2z$, $\sqrt{3}y^2z$, $\sqrt{3}y^2x$, $\sqrt{3}z^2x$, $\sqrt{3}z^2y$, and $\sqrt{6}xyz$ respectively, which are the components of a symmetric tensor of third rank. These quantities may also be viewed as products of x, y, z with the six α_i components, so as to give: $\beta_1 = x\alpha_1$; $\beta_2 = y\alpha_2$; $\beta_3 = z\alpha_3$; $\beta_4 = \sqrt{3}y\alpha_1 = \sqrt{(3/2)}x\alpha_6$; $\beta_5 = \sqrt{3}z\alpha_2 = \sqrt{(3/2)}y\alpha_6$; $\beta_6 = \sqrt{3}z\alpha_2 = \sqrt{(3/2)}y\alpha_4$; $\beta_7 = \sqrt{3}x\alpha_2 = \sqrt{(3/2)}y\alpha_6$; $\beta_8 = \sqrt{3}x\alpha_3 = \sqrt{(3/2)}z\alpha_5$; $\beta_9 = \sqrt{3}y\alpha_3 = \sqrt{(3/2)}z\alpha_4$; $\beta_{10} = \sqrt{3}x\alpha_4 = \sqrt{3}y\alpha_5 = \sqrt{3}z\alpha_6$.

4–3 The Symmetry-Coordinate Transformation (S-C-T) tables

In the preceding chapter, we have shown how the projection operator method can be used to convert hypervector components to symmetry coordinates, and have applied these considerations to components of polar vectors (denoted x, y, z), of axial vectors (denoted R_x, R_y, R_z), and of symmetric second-rank tensors (denoted x^2, y^2, $\ldots$, xy). We also showed how these symmetry coordinates are listed (in non-normalized form) alongside the various irreps on the right-hand side of conventional character tables (e.g. for C_6, C_{4v} and D_{3d} in Table 3–2). By listing the irrep to which each symmetry coordinate belongs, we are in effect stating how that coordinate transforms under the various symmetry operations, so that the information so given may be called *transformation tables*. It was pointed out, however, that the usual transformation tables attached to the character tables do not take the trouble to ensure that different sets of symmetry coordinates belonging to the same degenerate (E or T) irrep are similarly oriented. In Section 3–5, however, we have described how such similar orientation can be accomplished.

For the purposes of this book, such symmetry-coordinate transformation tables, with similarity of orientation included, are most valuable. Accordingly, we have given these transformation tables in Appendix E for all 32 crystallographic point groups, for polar and axial vectors and for symmetric second-rank and third-rank tensors. For the former two, we have used the conventional notation: x, y, z and R_x, R_y, R_z, respectively, for the components. The corresponding reps will be called Γ_{xyz} and $\Gamma_{\mathbf{R}}$ respectively. (Note that $\Gamma_{\mathbf{R}}$ was called Γ_{as} in Eq. (3–23).) For the symmetric second-rank tensors, however, we have used the six-vector notation α_1,

$\ldots, \alpha_6$ defined by Eq. (4). This permits listing of the Weyl-normalized components in a more compact notation than the usual $x^2, y^2, \ldots, \sqrt{2}xy$ notation. The corresponding rep will henceforth be denoted by $\Gamma_{\boldsymbol{\alpha}}$. Finally, the third-rank symmetric tensor components are also listed in Appendix E, as the set of components β_i, which generate a rep $\Gamma_{\boldsymbol{\beta}}$. The symmetry coordinates of these ten components and the irreps to which they belong were arrived at using the products of the set x, y, z with the $\alpha_1, \ldots, \alpha_6$ components as given by the relations presented below Eq. (5). By using products of the Γ_{xyz} and $\Gamma_{\boldsymbol{\alpha}}$ symmetry coordinates and the direct products of the associated irreps, we can generate the symmetry coordinates of $\Gamma_{\boldsymbol{\beta}}$ in a relatively straightforward way. Since these quantities are only required in the latter part of Chapter 9 (Section 9–5), however, most of our attention here will be given to the symmetry coordinates of Γ_{xyz}, $\Gamma_{\mathbf{R}}$, and $\Gamma_{\boldsymbol{\alpha}}$.

Whenever a degenerate irrep (E- or T-type) has more than one set of symmetry coordinates belonging to it, the pairs or triplets listed in Appendix E have been chosen so as to be of similar orientation. In the case of the uniaxial groups of types C_{nv}, D_n, D_{nh} (for $n = 3, 4, 6$), as well as D_{2d} and D_{3d}, we have always allowed the pair (α_4, α_5) to determine the orientation of coordinates belonging to the appropriate E irrep, rather than (x, y) or (R_x, R_y). This choice, first suggested by Nowick and Heller (1965), was made to simplify the handling of properties involving stress and strain variables, as we will see in later chapters. It is certainly permissible to do this, since any one of the pairs belonging to a given E irrep may be used to fix the orientation (see Section 3–5). In the case of the $\tilde{\text{E}}$ complex pair of one-dimensional irreps, there is no question of orientation, but it is still necessary to find the correct complex coordinates. Once the assignment of $x \mp iy$ has been identified, it is easy to obtain $\alpha_4 \pm i\alpha_5$ by noting that this quantity is proportional to $z(x \mp iy)$. (Recall that α_4 is related to yz and α_5 to xz.) Accordingly, by taking the product of the characters of the irrep of z and that of $x - iy$, we obtain the character of $\alpha_4 + i\alpha_5$ (similarly for the complex conjugate). In a similar way, note that $(\alpha_1 - \alpha_2) \mp i\alpha_6$ is proportional to $(x \mp iy)^2$, so that the correct irrep for it is obtained by squaring the characters of the irrep of $x \mp iy$. Problem 4–1 offers the reader an opportunity to utilize these short cuts for obtaining the irreps of the complex symmetry coordinates.

In these tables, it is interesting to note the distinction between the trigonal and hexagonal groups, which are sometimes difficult to make. (See, for example, the footnote on page 40, which points out why C_{3h} and

D_{3h} are regarded as hexagonal while S_6 is trigonal.) In the symmetry-coordinate transformation tables of Appendix E (henceforth referred to as the S-C-T tables) the distinction between these two crystal systems may be seen by the fact that, for the trigonal groups, the three pairs $(R_x, -R_y)$, (α_4, α_5) and $(\alpha_1 - \alpha_2, \alpha_6)$ all belong to the same E (or $\tilde{\mathrm{E}}$) irrep, while for the hexagonal groups they split up into two different E (or $\tilde{\mathrm{E}}$) irreps with $(R_x, -R_y)$ and (α_4, α_5) remaining together. This distinction will play an important role in determining the forms of the matter tensors for these two crystal systems.

Appendix E contains, in addition to the S-C-T tables for the 32 crystallographic point groups, two final tables for the infinite groups $C_{\infty v}$ and the full-rotation group called $R(3)$. The latter is an infinite group whose symmetry operations are all the rotations of a sphere (i.e. rotation by any angle and about any axis through the center). While it does not correspond to any crystal, it will be useful to us later when we wish to discuss the tensor properties of *isotropic* media, which possess such full rotation symmetry.

From these S-C-T tables it becomes possible to write down, almost immediately, the unitary matrix **S** that takes a hypervector **X** into symmetry coordinates, that is,

$$\overline{\mathbf{X}} = \mathbf{S}\mathbf{X} \tag{4–6}$$

In component from this becomes (see Section 3–4)

$$X^{(\gamma)}_{\mathrm{dr}} = S^{(\gamma)}_{\mathrm{dr},k} X_k \tag{4–6a}$$

where γ is the irrep, and d and r, the degeneracy and repeat indices respectively.

Since **S** is unitary, each row of coefficients (for a given triplet γ, d, r) is orthonormal to every other row. The symmetry coordinates, as listed in the tables, already provide the orthogonality; only the normalization needs to be added. Thus we must multiply all the coefficients of a given row by the factor needed to make

$$\sum_k |S^{(\gamma)}_{\mathrm{dr,k}}|^2 = 1 \tag{4–7}$$

We illustrate by giving the matrix **S** which transforms the hypervector $\boldsymbol{\alpha}$ to symmetry coordinates for the group C_{4v}. Appendix E shows that there are two symmetry coordinates belonging to irrep A_1: $\alpha_1 + \alpha_2$ and α_3, one belonging to B_1: $\alpha_1 - \alpha_2$, one to B_2: α_6, and a pair (α_4, α_5) to

E. Accordingly, inserting the normalization factors, the transformation matrix **S** is

$$\mathbf{S} = \begin{array}{l} A_1,1 \\ A_1,2 \\ B_1 \\ B_2 \\ E \\ \ \end{array} \begin{pmatrix} 1/\sqrt{2} & 1/\sqrt{2} & 0 & 0 & 0 & 0 \\ 0 & 0 & 1 & 0 & 0 & 0 \\ 1/\sqrt{2} & -1/\sqrt{2} & 0 & 0 & 0 & 0 \\ 0 & 0 & 0 & 0 & 0 & 1 \\ 0 & 0 & 0 & 1 & 0 & 0 \\ 0 & 0 & 0 & 0 & 1 & 0 \end{pmatrix}$$

Here the rows are labeled with irreps and, in the case of irrep A_1, with a repeat index. The columns, of course, correspond to $\alpha_1, \ldots, \alpha_6$ respectively.

4–4 The Fundamental Theorem

Consider a linear relationship between two physical quantities, both expressed as hypervectors. The Fundamental Theorem may then be expressed in two statements:

1. Only symmetry coordinates that belong to the same irrep and have the same degeneracy index can be coupled (i.e. related by a non-zero coefficient).
2. When symmetry coordinates of a degenerate irrep are coupled by a given coefficient, the same coefficient will also couple the partners of these symmetry coordinates (i.e. those having a different degeneracy index).

There is an important restriction, however: *the above statements only apply if different sets of symmetry coordinates that belong to the same degenerate irrep are similarly oriented.*

Because of the importance of this theorem to our subject* a rigorous proof is given in Appendix F, the basis of which is Neumann's principle. A good feeling for its reasonableness can be obtained, however, by considering a few cases. Suppose that physical quantities X and Y are related by $Y = KX$. Then K is a crystal property, which, by Neumann's principle, must be invariant under all symmetry operations. This can only be true if any operation that reverses the sign of X does the same for Y, that is, X and Y must transform in exactly the same way. Thus, if X is a symmetry coordinate that belongs to a given one-dimensional irrep (A or B), Y must belong to the same irrep.

For a two-dimensional (E-type) irrep, consider the case in which (X_1, X_2) and (Y_1, Y_2), which are two pairs of similarly oriented symmetry

* It is also basic to the subject of molecular vibrations (see Wilson *et al.*, 1955).

coordinates representing physical quantities, are linearly related. The requirements of Neumann's principle can only be met if

$$\left.\begin{aligned} Y_1 &= KX_1 \\ Y_2 &= KX_2 \end{aligned}\right\} \tag{4–8}$$

with the same K in both equations. Equations (8) allow not only for operations that change the sign of X_1 or X_2 (as in the previous one-dimensional case) but also for operations that take X_1 into X_2 and Y_1 into Y_2 (or into linear combinations). Note that Eqs. (8) have utilized both statements (1) and (2) of the fundamental theorem. From (1), X_1 can only couple to Y_1 and X_2 to Y_2, and from (2) the quantity K in both equations must be the same.

Then, when can more than one symmetry coordinate be involved in a linear relationship, so as to produce off-diagonal (cross) terms? Answer: only if an irrep is *repeated*. Thus, suppose that X_1 and X_2 both belong to a given one-dimensional irrep and so does Y_1. Then the relation

$$Y_1 = K_1X_1 + K_2X_2$$

is allowed by Neumann's principle as well as by our Fundamental Theorem. This argument may be similarly extended to degenerate irreps.

The ultimate case of repeated irreps occurs for a triclinic crystal belonging to the group C_1 (no symmetry at all). Here there is only the totally symmetric irrep, called A, to which all coordinates (all of which are, therefore, symmetry coordinates) belong. In this case all coordinates are repeats of the A irrep, and, therefore, all are coupled. Of course, there is no simplification here due to crystal symmetry, because there is no symmetry.

In the opposite extreme, an important consequence of the Fundamental Theorem is that $\mathbf{K}$ is identically zero if and only if $\Gamma_\mathbf{X}$ and $\Gamma_\mathbf{Y}$ (the reps generated by $\mathbf{X}$ and $\mathbf{Y}$, respectively) have no irreps in common.

4–5 Applications of the Fundamental Theorem

We now apply the Fundamental Theorem to the linear relation (1) with $\mathbf{X}$ and $\mathbf{Y}$ in hypervector form (with dimensionality n_1 and n_2 respectively). First we transform to symmetry coordinates with

$$\overline{\mathbf{X}} = \mathbf{SX}; \quad \overline{\mathbf{Y}} = \mathbf{TY} \tag{4–9}$$

where $\mathbf{S}$ and $\mathbf{T}$ are both unitary. Correspondingly, the physical property matrix $\mathbf{K}$ is transformed into

$$\overline{\mathbf{K}} = \mathbf{TKS}^\dagger \tag{4–10}$$

so that

$$\overline{\mathbf{Y}} = \overline{\mathbf{KX}} \tag{4–11}$$

As described in Section 3–4, the components of $\overline{\mathbf{Y}}$ and $\overline{\mathbf{X}}$ are each labeled with a triplet of indices γ, d, r for the irrep, degeneracy index and repeat index, respectively, for example $Y^{(\gamma)}_{\mathrm{dr}}$. Correspondingly, it may be expected that the components of $\overline{\mathbf{K}}$ should be labeled by two sets of triplets; however, the Fundamental Theorem allows a substantial simplification of $\overline{\mathbf{K}}$ to the form given here:

$$Y^{(\gamma)}_{\mathrm{dr}} = K_{\gamma,\mathrm{rs}} X^{(\gamma)}_{\mathrm{ds}} \tag{4–12}$$

in which the right-hand side is summed only over the repeat index s. The reason is that the theorem requires that we have the same γ and d on both sides of the equation, and, further, that the component of $\overline{\mathbf{K}}$ must be independent of degeneracy index d. The reason for placing γ differently on K than on Y and X is the following. The quantities Y and X belong to, that is, transform as, the irrep γ. On the other hand $\overline{\mathbf{K}}$, which is a physical property, must be *invariant* (by Neumann's principle). It is labeled γ only to denote that it couples quantities that transform as γ. (This is a very important point, which the reader must make sure he/she understands.)

It is now relatively straightforward to determine the number of independent coefficients that appear in the matrix $\overline{\mathbf{K}}$ (and, therefore, in $\mathbf{K}$ as well). Consider, first, the case in which γ is a one-dimensional (non-degenerate) irrep (A or B) and where there are no repeats in either $Y^{(\gamma)}$ or $X^{(\gamma)}$. Then, Eq. (12) becomes simply

$$Y^{(\gamma)} = K_\gamma X^{(\gamma)} \tag{4–13}$$

so that just one coefficient is contributed by this relationship. On the other hand, for a one-dimensional irrep in which a repeat exists either in $\overline{Y}$ or in $\overline{X}$ we obtain*

$$Y^{(\gamma)} = K_{\gamma,1} X^{(\gamma)}_{01} + K_{\gamma,2} X^{(\gamma)}_{02} \tag{4–14}$$

when the repeat is in $\overline{\mathbf{X}}$, or

$$Y^{(\gamma)}_{01} = K_{\gamma,1} X^{(\gamma)}; \quad Y^{(\gamma)}_{02} = K_{\gamma,2} X^{(\gamma)} \tag{4–15}$$

when the repeat is in $\overline{\mathbf{Y}}$. (The zero in the subscript of $X^{(\gamma)}_{01}$ or $Y^{(\gamma)}_{01}$ is used to denote the absence of a degeneracy index.) Thus, in either case there are two independent coefficients. If both $\overline{\mathbf{Y}}$ and $\overline{\mathbf{X}}$ involve two coordinates belonging to the non-degenerate irrep γ, the result is clearly

* Clearly, this case only occurs if **X** and **Y** have different tensor character.

$$\left.\begin{aligned} Y_{01}^{(\gamma)} &= K_{\gamma,11}X_{01}^{(\gamma)} + K_{\gamma,12}X_{02}^{(\gamma)} \\ Y_{02}^{(\gamma)} &= K_{\gamma,21}X_{01}^{(\gamma)} + K_{\gamma,22}X_{02}^{(\gamma)} \end{aligned}\right\} \tag{4–16}$$

so that four independent K_γ coefficients are introduced. On the other hand, if **K** is intrinsically a symmetric matrix, its symmetry is retained under the unitary transformation to symmetry coordinates (see Problem 4–5), and we must then have

$$K_{\gamma,12} = K_{\gamma,21} \tag{4–17}$$

In this case, then, there are only three independent K_γ coefficients. Extension of the above arguments to a larger number of repeats in either $\bar{\mathbf{X}}$ or $\bar{\mathbf{Y}}$ or both is now easily made and the results are given in Table 4–1. Note that **K** can be symmetric only if **Y** and **X** are similar hypervector quantities, so that $n_1 = n_2$ and **K** is a square matrix in such cases.

Next we consider the case of a two-dimensional irrep γ. Since, from the fundamental theorem, there can be no K_γ coefficients which relate different degeneracy indices, d, and since the corresponding coefficients for different d's are the same, we must have the same results as for non-degenerate irreps. For example, consider the case of a single pair $(Y_1^{(\gamma)}, Y_2^{(\gamma)})$ for $\bar{\mathbf{Y}}$ and a repeated pair $(X_{11}^{(\gamma)}, X_{21}^{(\gamma)})$ and $(X_{12}^{(\gamma)}, X_{22}^{(\gamma)})$ for $\bar{\mathbf{X}}$, analogous to the case of Eq. (14). In this case, we have

$$\left.\begin{aligned} Y_1^{(\gamma)} &= K_{\gamma,1}X_{11}^{(\gamma)} + K_{\gamma,2}X_{12}^{(\gamma)} \\ Y_2^{(\gamma)} &= K_{\gamma,1}X_{21}^{(\gamma)} + K_{\gamma,2}X_{22}^{(\gamma)} \end{aligned}\right\} \tag{4–18}$$

so that there are still only two independent coefficients of type K_γ.

The above argument extends equally well to three-dimensional (T-type) irreps. Thus, the upper part of Table 4–1 giving the numbers of coefficients, either when **K** is intrinsically symmetric or not, applies equally well to degenerate as to non-degenerate irreps.

The case of complex pairs of irreps (of type $\tilde{\mathrm{E}}$) gives different results, however. Because the characters of the two non-degenerate irreps are complex conjugates of each other, application of the projection operator method (Eq. (3–16)) ensures that the corresponding symmetry coordinates must be complex conjugates of each other. If we call the pair of irreps γ_1 and γ_2, respectively, and first consider the case in which there are no repeats in either $\bar{Y}$ or $\bar{X}$, the symmetry coordinates may be of the form $X^{(\gamma_1)} = X_1 - \mathrm{i}X_2$ and $X^{(\gamma_2)} = X_1 + \mathrm{i}X_2$, and similarly for the $\bar{Y}$ coordinates. Then applying the Fundamental Theorem to irrep γ_1:

$$Y_1 - \mathrm{i}Y_2 = K_{\tilde{\mathrm{E}}}(X_1 - \mathrm{i}X_2) \tag{4–19}$$

where, in general, $K_{\tilde{\mathrm{E}}}$ is a complex quantity. The equation for irrep γ_2 is obtained by taking the complex conjugate of Eq. (19):

$$Y_1 + iY_2 = K_{\tilde{E}}^{*}(X_1 + iX_2) \tag{4–20}$$

since $X^{(\gamma_2)} = X^{(\gamma_1)*}$ and $Y^{(\gamma_2)} = Y^{(\gamma_1)*}$. Thus, the coefficient that relates $Y^{(\gamma_2)}$ to $X^{(\gamma_2)}$ is not an independent quantity but is just the complex conjugate of that for γ_1. (This is the reason that we label the constant $\tilde{E}$ instead of γ_1 and γ_2.) From Eqs. (19) and (20), and expressing the complex quantity $K_{\tilde{E}}$ in terms of its real and imaginary parts:

$$K_{\tilde{E}} = K_{\tilde{E}}^{re} + iK_{\tilde{E}}^{im} \tag{4–21}$$

where $K_{\tilde{E}}^{re}$ and $K_{\tilde{E}}^{im}$ are both real, we conclude that there are *two* independent (real) coefficients involved in this case. By adding Eqs. (19) and (20) and utilizing (21), we obtain

$$Y_1 = K_{\tilde{E}}^{re}X_1 + K_{\tilde{E}}^{im}X_2 \tag{4–22}$$

and by subtracting Eq. (20) from (19):

$$Y_2 = -K_{\tilde{E}}^{im}X_1 + K_{\tilde{E}}^{re}X_2 \tag{4–23}$$

Equations (22) and (23) involve only real quantities and are, therefore, more useful than (19) and (20) in relation to measurements. The quantities $K_{\tilde{E}}^{re}$ and $K_{\tilde{E}}^{im}$ are then the coefficients of the physical property matrix that are of interest. Thus, if we deal with the real quantities (X_1, X_2), (Y_1, Y_2) and treat these as the symmetry coordinates, the block of the $\bar{\mathbf{K}}$ matrix associated with the pair of irreps $\tilde{E}$ may be written as

$$\begin{pmatrix} K_{\tilde{E}}^{re} & K_{\tilde{E}}^{im} \\ -K_{\tilde{E}}^{im} & K_{\tilde{E}}^{re} \end{pmatrix} \tag{4–24}$$

In the case where $\mathbf{K}$ is intrinsically symmetric, however, $K_{\tilde{E}}^{im}$ must be zero (in order to retain the symmetry in Eqs. (22) and (23)). In that case there is only one independent constant.

The situation differs if the matching complex symmetry coordinates are $X_1 \mp iX_2$ and $Y_1 \mp iY_2$ for the γ_1 and γ_2 irreps of $\tilde{E}$, respectively, that is, with the signs for X and Y oppositely matched. In that case the matrix block is readily shown to take the form

$$\begin{pmatrix} K_{\tilde{E}}^{re} & K_{\tilde{E}}^{im} \\ K_{\tilde{E}}^{im} & -K_{\tilde{E}}^{re} \end{pmatrix} \tag{4–25}$$

which involves two independent constants. (The question of whether $\mathbf{K}$ is intrinsically symmetric does not arise here, since the occurrence of opposing signs of the complex quantities means that $\mathbf{X}$ and $\mathbf{Y}$ are not similar hypervector quantities.)

For an $\tilde{E}$ irrep in which there is one $Y^{(\tilde{E})}$ and two $X^{(\tilde{E})}$ (say $(X_1 \mp iX_2)$ and $(X_3 \mp iX_4)$), we obtain for the first $\tilde{E}$ irrep

$$Y_1 - \mathrm{i}Y_2 = K_{\tilde{E},1}(X_1 - \mathrm{i}X_2) + K_{\tilde{E},2}(X_3 - \mathrm{i}X_4)$$

and, since both $K_{\tilde{E},1}$ and $K_{\tilde{E},2}$ are complex quantities there are four independent real constants. The matrix block of the Ẽ irrep, involving the real quantities, is

$$\begin{pmatrix} K_{\tilde{E},1}^{re} & K_{\tilde{E},1}^{im} & K_{\tilde{E},2}^{re} & K_{\tilde{E},2}^{im} \\ -K_{\tilde{E},1}^{im} & K_{\tilde{E},1}^{re} & -K_{\tilde{E},2}^{im} & K_{\tilde{E},2}^{re} \end{pmatrix} \tag{4–26}$$

A similar result, but with a transposed matrix, is obtained for two $Y^{(\tilde{E})}$ and one $X^{(\tilde{E})}$. Finally, if both $X^{(\tilde{E})}$ and $Y^{(\tilde{E})}$ are repeated, we readily see that eight independent real constants are required, but in the case where **K** is symmetric this reduces to only four. These results are listed in the bottom half of Table 4–1. In summary, a complex Ẽ irrep gives results that differ both from the case of a two-dimensional E irrep and from that of two independent irreps.

The total number of independent coefficients **K** is simply obtained by adding up the numbers for all irreps that are common to both **X** and **Y**. For example, consider the case in which **Y** is a vector and **X** a symmetric tensor of second rank. We must then look for common irreps between Γ_{xyz} and Γ_{α} (which we previously called Γ_S). Let us take the cases of two trigonal crystal symmetries. First, for C_{3v}, we see from Appendix E that under irrep A_1 there is one symmetry coordinate of xyz and two of α. From Table 4–1, matching of these gives rise to two constants. Similarly, for the E irrep, the coordinates are again 1:2, giving two more constants. Thus, the total number of constants is four. Second, we turn to group C_3. Again, the A irrep gives rise to two constants, but the 1:2 matching in the Ẽ irreps gives rise to four additional constants, leading to a total of six

Table 4–1. *Number of independent coefficients of a matter tensor **K** coming from each irrep*

Irrep	No. of repeats for $Y{:}X$	No. of coeffs **K** non-symm.	**K** symm.
A, B, E, or T	1:1	1	1
	1:2 (or 2:1)	2	—
	2:2	4	3
	2:3 (or 3:2)	6	—
	3:3	9	6
Ẽ	1:1	2	1
	1:2 (or 2:1)	4	—
	2:2	8	4

independent constants. (Note that here there is no question of **K** being symmetric since **X** and **Y** are not tensors (or hypervectors) of the same type.)

In this way, the material in Table 4–1, together with the S-C-T tables, will be useful in allowing us to determine the total number of coefficients involved in any physical property matrix **K** for any given crystal structure. The considerations of this section, however, have given us substantially more information than just the numbers of independent coefficients. Specifically, they allow us to determine which are the independent coefficients, that is, where they are located in the matrix $\bar{\mathbf{K}}$ of Eq. (11). Reversing the transformation of Eq. (10), then, enables us to determine the form of the original matrix **K**. This, in fact, will be the major objective of the latter part of this book (Chapters 6–10); the group theoretical groundwork for this objective has been laid in the present chapter.

4–6 Alternative treatments

In this closing section, we shall review two alternative ways that have been used to approach the problem of determining the form of **K** under a given group of symmetry operations.

The first method is concerned with determining the effect of a symmetry operation, R, on the form of the matter tensor **K** in Eq. (1). The effect of R on tensors **X** and **Y** is to transform them to quantities **X**′ and **Y**′ given, respectively, by

$$\left.\begin{aligned} \mathbf{X}' &= \mathbf{D}_{\mathbf{X}}(R)\mathbf{X} \\ \mathbf{Y}' &= \mathbf{D}_{\mathbf{Y}}(R)\mathbf{Y} \end{aligned}\right\} \qquad (4\text{–}27)$$

where the set of matrices $\mathbf{D}_{\mathbf{X}}(R)$ is just the group representation generated by **X** as a basis, and similarly for $\mathbf{D}_{\mathbf{Y}}(R)$. These are unitary matrices, so that

$$\mathbf{D}(R)^{-1} = \mathbf{D}(R)^{\dagger} \qquad (4\text{–}28)$$

If, now, Eqs. (27) are inverted (i.e. solved for **X** and **Y**, respectively) and inserted into Eq. (1), we obtain

$$\mathbf{D}_{\mathbf{Y}}{}^{\dagger}\mathbf{Y}' = \mathbf{K}\mathbf{D}_{\mathbf{X}}{}^{\dagger}\mathbf{X}'$$

or (4–29)

$$\mathbf{Y}' = (\mathbf{D}_{\mathbf{Y}}\mathbf{K}\mathbf{D}_{\mathbf{X}}{}^{\dagger})\mathbf{X}' = \mathbf{K}'\mathbf{X}'$$

which expresses the relation between **Y** and **X** in the new coordinate system following the symmetry operation, and defines the matter tensor **K**′. However, Neumann's principle requires that $\mathbf{K}' \equiv \mathbf{K}$. Accordingly, as a consequence of the symmetry operation R, we obtain the following equation for $\mathbf{K}$ from Eq. (29):

$$\mathbf{D}_{\mathbf{Y}}(R)\mathbf{K}\mathbf{D}_{\mathbf{X}}(R)^{\dagger} = \mathbf{K} \tag{4–30}$$

For a group of order h, there will be $h - 1$ such relations (omitting the identity operation), all of which can serve to simplify the structure of $\mathbf{K}$. (Actually only operations R which are the generators of the group give independent information on the form of an equation of type (30), so that there are fewer than $h - 1$ independent relations.) A particularly simple version of this approach was developed by Fumi (1952a) as the 'direct inspection method'; however, the method only works easily for symmetry groups of one-, two- and four-fold principal axes, that is, those for which the Cartesian coordinates do not transform into linear combinations of themselves under the symmetry operations.

The use of Eq. (30) provides the most conventional way (Nye, 1957) to simplify a matter tensor in accordance with crystal symmetry. It is more awkward to apply than the method presented in this chapter, since it does not take advantage of the power of group representation theory. Nevertheless, Eq. (30) will be useful to us in Chapter 5 for determining what happens to $\mathbf{K}$ under operations that involve time reversal.

The second alternative method is more sophisticated and does make use of representation theory (Fumi, 1952b; Juretschke, 1974). It starts directly from the matter tensor $\mathbf{K}$ without reference to the field tensors $\mathbf{X}$ and $\mathbf{Y}$. In this method, one determines the transformation properties of the tensor $\mathbf{K}$ under the symmetry operations of the group in the same way as was done for the field quantities $\mathbf{X}$ and $\mathbf{Y}$ earlier in this chapter, that is, by obtaining symmetry coordinates. However, $\mathbf{K}$ is a matter tensor, and, by Neumann's principle, must be invariant under all the group operations. Therefore, the only non-zero symmetry coordinates of $\mathbf{K}$ are those that transform as A_1, the totally symmetric irrep.

The number of independent constants is the number of times that A_1 appears in the reduction of $\Gamma_{\mathbf{K}}$, where $\Gamma_{\mathbf{K}}$ is the reducible rep formed by the transformation matrices of $\mathbf{K}$. It is given by Eq. (3–12) with $\chi^{(\alpha)}(R)^* = 1$ for every R (since α is now A_1) and $\chi(R)$ is the character of the rep $\Gamma_{\mathbf{K}}$ for symmetry operation R.

While this method is conceptually simple, it becomes difficult to apply when the tensor $\mathbf{K}$ is of high rank (four or greater), especially for those groups in which the Cartesian orthogonal coordinates transform into linear combinations of themselves under symmetry operations of the crystal. Accordingly, the method is more often used to obtain the number of independent constants of **K** rather than their actual forms. The method of symmetry coordinates used in this book (Section 4–5), by working with the lower-rank field tensors **X** and **Y**, turns out to be much easier to apply.

As a direct consequence of either of these two alternative methods, however, we have an important theorem which may be stated as follows:

The form of a given matter tensor in a given crystal class depends only on its tensor character and its intrinsic symmetry, but not on the quantities that it relates. (By the 'form' of **K** we mean the coefficients that are zero as well as the interrelationships among the non-vanishing coefficients.)

This theorem may be applied to the matter tensors that appear in Tables 1–1 and 2–1. Thus, for example, thermal expansion α_{ij} relates a second-rank symmetric tensor and a scalar, while dielectric permeability κ_{ij} relates two polar vectors; nevertheless, since both are second-rank symmetric matter tensors, their structures must be the same in any given point group. This conclusion will be useful to us in subsequent chapters of this book, since it will enable us to handle widely different crystal properties using the same treatment.

Problems

4–1. With the aid of the character table (Table 3–2), work out the S-C-T table for the components of a polar vector (x, y, z), an axial vector (R_x, R_y, R_z) and a second-rank symmetric tensor (α_i) for the uniaxial point groups C_{4h} and S_6.

4–2. Do the same as Problem 4–1 for the uniaxial groups C_{6v}, D_4 and D_{3d}, including the similarity of orientations for components belonging to the E irreps, using the methods of Section 3–5.

4–3. Do the same for the cubic group T_d, including similarity of orientation for components belonging to the T_2 irrep.

4–4. Write the three unitary matrices **S** (Eq. (6)) that transform a polar vector (x, y, z), an axial vector (R_x, R_y, R_z) and the hypervector $\boldsymbol{\alpha}$ $(\alpha_1, \ldots, \alpha_6)$ for the group D_{3d} using the S-C-T table.

4–5. Show that, if **K** is symmetric and transforms according to Eq. (10) with $\mathbf{S} = \mathbf{T}$ (**X** and **Y** being similar hypervectors), the similarity transformed $\bar{\mathbf{K}}$ is also symmetric.

4–6. Note the statement below Eq. (26) that 'a complex $\tilde{\mathrm{E}}$ irrep gives results that differ both from the case of a two-dimensional E irrep and from that of two independent one-dimensional irreps'. To verify this statement, compare the matrix of Eq. (26) with the corresponding ones for these other two cases.

4–7. From Table 4–1, obtain the total number of independent constants for the case where **Y** is a vector and **X** a symmetric second-rank tensor, for the groups D_4, S_6 and T_d.

4–8. Do the same for the case of **X** and **Y** both symmetric second-rank tensors for the groups C_{4v} and D_3.

4–9. Work out the S-C-T table for the symmetry coordinates of the ten triple products β_i of vector components for point group D_4. (Take the β_i's as products of x, y, z with the six α_i components, making use of the direct products of the appropriate irreps.) Be sure that the symmetry coordinates belonging to irrep E are oriented similarly to (α_4, α_5).

4–10. Work out the S-C-T table β_i coordinates for point groups C_6 and C_{6v}.

5

The magnetic point groups and time reversal

While X-ray diffraction reveals the spatial arrangement of the atoms of a crystal, diffraction by neutrons allows the additional determination of the type and degree of order of atomic magnetic moments (spins) in a crystal (the magnetic symmetry). This is possible because neutrons possess a net magnetic moment, so that they interact with the magnetic moments of the atoms of the crystal and, therefore, can distinguish an atom with spin in one direction from one with spin in a different direction.

Magnetic crystals include ferromagnetic, ferrimagnetic and antiferromagnetic types, shown schematically in Fig. 5–1. The symmetry of such crystals is not solely described by the spatial arrangement of the atoms, but includes, in addition, the orientation of atomic magnetic moments. In a ferromagnetic crystal, the magnetic spins are arranged preferentially parallel to one particular crystallographic direction. Accordingly, it can possess a magnetic moment even in the absence of an applied magnetic

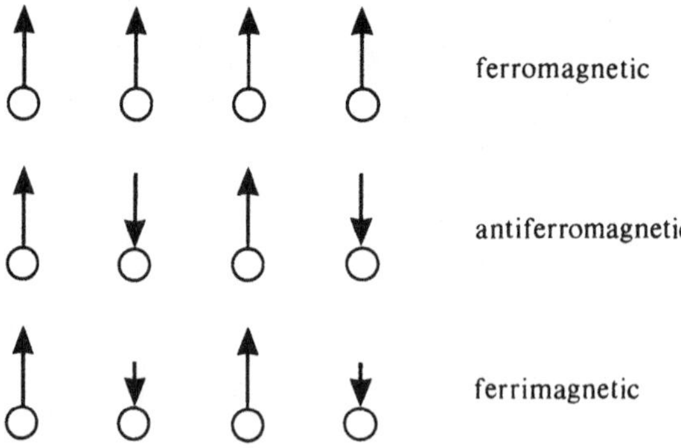

Fig. 5–1. Schematic illustration of the spin arrangements for three basic types of magnetic structures.

field. Examples are the metals: iron, nickel and cobalt. In the antiferromagnetic case, the spins are ordered in an antiparallel arrangement, so that the crystal has a zero net moment. Examples are the metal chromium and the oxide MnO. Finally, for ferrimagnetism there are two unequal spins aligned antiparallel to each other, so that the crystal possesses a net magnetic moment. Notable examples among the oxides are the ferrites and the garnets.

Reversal of spin may be considered as a new symmetry operation, θ. If we think of magnetic moments as generated by currents, then reversal of magnetic moments corresponds to reversal of current direction, that is, to *time reversal*. With the introduction of θ, we find that there are many more symmetry possibilities than those given by the 32 classical point groups. The subject of magnetic symmetry was pioneered by several Soviet theoreticians, notably A. V. Shubnikov (see Shubnikov and Belov, 1964).

5–1 The magnetic point groups

With time reversal, θ, as a symmetry operation, we can have θ appearing explicitly, or as an ordinary symmetry operation R followed by (or preceded by) θ, that is, θR. Note that, since θ and R are different types of operation, they inevitably commute, that is, $\theta R = R\theta$. Such products comprise a new type of operation called a *complementary operation*, which for brevity can be denoted by R with a bar underneath it, namely $\underline{R}$. With the aid of complementary operations, point groups may be formed that are conveniently divided into the following three categories:

I. Groups that do not include θ, either explicitly or in complementary operations. These are, then, the 32 classical point groups.

II. Groups that include θ explicitly as well as in the form of complementary operations. By the requirement of closure, such a group, $\mathscr{G}'$, must have twice as many elements as the corresponding classical point group, $\mathscr{G}$, from which it was derived. (Specifically, for every element R in $\mathscr{G}$, there are the elements R and θR in $\mathscr{G}'$.) Accordingly, there are 32 groups of type II.

III. Groups that do not include θ explicitly but have at least one complementary operation. We will see that, in fact, in such groups exactly half of the operations are of the classical type while the other half are complementary operations. We will also see that there are 58 variants of the crystallographic point groups in this category.

Shubnikov regarded an atomic species with spin as black or white, depending on whether the spin orientation is 'up' or 'down', and the spin reversal operation as one that changes color from black to white, or vice versa. In these terms, the type-I groups, which apply to crystals in which the orientation of all spins remains invariant under all spatial point-symmetry operations, may be considered *singly* colored groups (either black or white). The type-II groups, in which θ appears explicitly, cannot be used to describe magnetic structures and in fact can only describe non-magnetic (dia- or paramagnetic) crystals which are time symmetric.* In the color interpretation of Shubnikov, these are called the *gray groups*, since no atomic site is either black or white.

Finally, the most interesting new additions to our list of point groups come from the type-III groups. In Shubnikov's terms these are the *doubly colored* or *black-and-white* groups. They may be formed as follows (for a proof, see Appendix G). Given a classical point group $\mathscr{G}$ of order g, choose a subgroup $\mathscr{H}$ of $\mathscr{G}$ whose order is $g/2$. Then multiply by θ all the elements of $\mathscr{G} - \mathscr{H}$ (i.e. elements of $\mathscr{G}$ not in $\mathscr{H}$). If $\mathscr{H} = \{A_i\}$ (i.e. the set A_i) and $\mathscr{G} - \mathscr{H} = \{B_i\}$, then the new magnetic group $\mathscr{M} = \mathscr{H} + \theta(\mathscr{G} - \mathscr{H})$. Stated differently, $\mathscr{M}$ is made up of elements $\{A_i\} + \{\underline{B}_i\}$. In this way, one obtains 58 variants of the 32 classical point groups which constitute all of the groups of type III.

As already mentioned, groups of type II are not consistent with magnetic structures of crystals. Thus only types I and III are suitable to magnetic crystals. These two types taken together constitute the 90 *magnetic point groups*, and are presented in Table 5–1 at the end of this chapter, page 90–1. In this table, for each type-I group the symmetry operations are simply listed by classes. On the other hand, for the related variant (type-III) groups, + and − signs are shown denoting, respectively, which operations in the variant group appear unaltered, and which appear in complementary form. Note that in all cases the number of operations with + signs and − signs are equal, in accordance with the method of forming the type-III groups described above. The nomenclature used for the type-III magnetic groups is of the form: $\mathscr{G}$: $\mathscr{H}$, that is, with the subgroup $\mathscr{H}$ given after a colon.† The final column of Table 5–1, listing irreps, will be explained later.

Another method for obtaining the black-and-white groups is based on

* Some antiferromagnetic structures do belong to these non-magnetic classes, however (Birss, 1964, p. 82). This can occur when the product θt is a symmetry operation, where t is a translation by less than a lattice distance. An example is that of antiferromagnetic chromium metal. Since the macroscopic properties of such a crystal do not depend on the translational component t, θ must be admitted as a symmetry operation in this case.

† The International notation for the magnetic groups is confusing for our purposes, but may be found in standard references, for example Birss (1964).

the character table of the corresponding classical group $\mathcal{G}$. One looks at each of the one-dimensional real irreps of $\mathcal{G}$ for which all characters are ± 1, that is, all except the A_1, irrep. For *each* such irrep, one can convert the + and − signs of the characters into the + and − signs of Table 5–1 and thereby obtain a suitable type-III group. This method then avoids the need to search for all possible subgroups $\mathcal{H}$ of order $g/2$ and allows the magnetic point groups (Table 5–1) to be obtained directly from the character tables (Table 3–2). The proof of this method is given in Appendix G as a by-product to the main objective of that appendix, which will be discussed later in this chapter.

5–2 Neumann's principle in space-time

We again turn to our basic relation

$$\mathbf{Y} = \mathbf{K}\mathbf{X} \tag{5–1}$$

in which a matter tensor **K** relates two physical quantities **X** and **Y**. (The latter quantities may be either in tensor form or expressed as hypervectors.) Time reversal may affect the quantities **X** and **Y**. Thus, a magnetic field **H** or magnetic induction **B** may be regarded as derived from currents and, therefore, should reverse its sign upon time reversal:

$$\mathbf{H}(-t) = -\mathbf{H}(t) \tag{5–2}$$

Most physical quantities, on the other hand, would be unaffected by time reversal. A commonly used terminology is to refer to all tensors that are invariant under θ as *i-tensors*, and those that reverse sign under θ (as, for example, **H** in Eq. (2)) as *c-tensors*. This terminology may be applied both to the physical quantities (here, **X** and **Y**) and to the matter tensors (here, **K**). Under time reversal, the matrix **K** describing the interaction between **X** and **Y** must be compatible with the symmetry of these two quantities as well as that of the crystal. We may then restate Neumann's principle in a form which is a generalization of that given in Chapter 4, as follows:

The symmetry of a physical property tensor **K** *must include all the symmetry of the crystal in space-time.*

There is, however, an important limitation to this principle, namely, that it is confined to phenomena of macroscopically reversible thermodynamics, that is, to equilibrium properties of crystals, dealt with in Chapter 1. It is not valid for the case of transport properties (see Chapter 2), where the system is permanently in non-equilibrium but has reached a steady state. The reason for this distinction between the two classes of property is related to the second law of thermodynamics, which may be stated as: in any process which occurs for an isolated system, the entropy, s, of the system never decreases. In the case of a system in equilibrium, reversible changes take place for which $\Delta s = 0$; in such a case the two directions of time are equivalent, and time reversal is permissible. In the case of a system undergoing an irreversible change, as in a transport process, $\Delta s > 0$, or entropy is being produced. In such systems, there *is* an intrinsically preferred direction of time describing the phenomenon, which cannot be reversed; thus time reversal is not a permissible symmetry operation of the crystal.

An example shows how profound this conclusion actually is. Consider electric transport, described by the equation

$$\mathbf{j} = \boldsymbol{\sigma}\mathbf{E} \tag{5-3}$$

in which $\mathbf{j}$ is the current density, $\mathbf{E}$ the electric field and $\boldsymbol{\sigma}$ the conductivity of the material. Under time inversion $\mathbf{j}$ is reversed while $\mathbf{E}$ is unchanged; therefore, σ is time antisymmetric, that is, $\sigma(-t) = -\sigma(t)$. Accordingly, if for non-magnetic crystals (type II), the time reversal operation θ were applicable, we would obtain: $\boldsymbol{\sigma} = -\boldsymbol{\sigma}$. This would mean that a time-antisymmetric property like $\boldsymbol{\sigma}$ must vanish identically for such crystals. Of course, this is not the case, since non-magnetic crystals do indeed possess finite conductivities.

We conclude that the generalized Neumann principle in space-time is applicable only to static (equilibrium) phenomena and not to dynamic (transport) phenomena. In the case of transport phenomena, time reversal or any (complementary) operation that includes time reversal is not permissible, and we therefore deal only with classical symmetry operations. In the case of type-I groups $\mathcal{G}$, containing only classical operations, this offers no problem. In the case of type-II groups $\mathcal{G}'$, containing $\mathcal{G}$ as well as θ and all complementary operations of $\mathcal{G}$, only the operations of group $\mathcal{G}$ are applicable. Finally, for the type-III (black-and-white) groups $\mathcal{M}$, only the classical operations contained in subgroup $\mathcal{H}$ are applicable. In the case of types-I and -II groups, therefore, the corresponding matter tensor will be that appropriate to a crystal belonging to the classical point group $\mathcal{G}$. In the case of the type-III group, matter tensors will be those appropriate to a crystal belonging to the classical point group $\mathcal{H}$. These remarks apply specifically to transport phenomena.

The rest of this chapter is, therefore, devoted to the consequences of the generalized Neumann principle for the case of static (equilibrium) phenomena.

5–3 Application to non-magnetic crystals

In this section we apply Neumann's principle in space-time to the properties of non-magnetic crystals, that is, to the gray groups $\mathcal{G}'$ of type II which contain θ explicitly. If $\mathcal{G} = \{R_i\}$ is the corresponding classical point group of order g, then there are $2g$ elements in $\mathcal{G}'$, namely, the sets $\{R_i\}$ and $\{\theta R_i\}$.

Let us first consider the case in which both **Y** and **X** are *i*-tensors (i.e. time symmetric). We then obtain g relations of the type of Eq. (4–30), namely,

$$\mathbf{K} = \mathbf{D_Y}(R)\mathbf{K}\mathbf{D_X}(R)^\dagger$$

from all operations R_i and an *identical* g relation from operations θR_i. Therefore, the results for the structure of **K** are the same as if θ had not been present at all, that is, as if they had been obtained just from the classical point group $\mathcal{G}$.

Next, we consider what happens if both **X** and **Y** are *c*-tensors (time antisymmetric). Again, exactly the same result is obtained, namely, the set of g relations obtained from operations $\{\theta R_i\}$ are the same as those obtained from $\{R_i\}$. These two cases have in common the fact that **K** is an *i*-tensor in both cases; for, if both **X** and **Y** are time antisymmetric **K** will be time symmetric.

The remaining case to be considered is that for which **Y** is an *i*-tensor and **X** a *c*-tensor, or vice versa. In either case **K** is now a *c*-tensor. Here, for operation R, we obtain Eq. (4–30), while for its complementary operation θR, or $\underline{R}$, we obtain

$$\mathbf{K} = -\mathbf{D_Y}(R)\mathbf{K}\mathbf{D_X}(R)^\dagger$$

This means that $\mathbf{K} = -\mathbf{K}$, which can be valid if **K** is identically zero.

We conclude, therefore, that for non-magnetic crystals, if **K** is an *i*-tensor, its structure is the same as that obtained from the corresponding classical group $\mathcal{G}$. On the other hand, if **K** is a *c*-tensor, it must be identically zero. The latter conclusion is strikingly different from that which would have been obtained if time reversal had not been taken into consideration. As an example, consider the magnetoelectric polarization tensor, λ_{ij}, of Eq. (1–10) and Table 1–1. This is clearly a *c*-tensor and, therefore, must vanish identically for any non-magnetic crystal.

5–4 Application to magnetic crystals

In the case of magnetic crystals belonging to the black-and-white (type-III) groups, it is advantageous to consider simultaneously the magnetic group $\mathcal{M}$ and its corresponding classical point group $\mathcal{G}$. As pointed out in Section 5–1,

$$\mathcal{G} = \{A_i\} + \{B_i\} = \mathcal{H} + (\mathcal{G} - \mathcal{H}) \tag{5-4}$$

while

$$\mathcal{M} = \{A_i\} + \{\underline{B}_i\} = \mathcal{H} + \theta(\mathcal{G} - \mathcal{H}) \tag{5-5}$$

where $\{A_i\}$ is the set of operations of the subgroup $\mathcal{H}$.

Consider, first, the case in which **Y** and **X** are both *i*-tensors in $\mathcal{M}$. Then **Y** and **X** transform in exactly the same way in groups $\mathcal{G}$ and $\mathcal{M}$, and relations of the type of Eq. (4–30) are the same in both groups. Accordingly, the structure of **K** is identical in groups $\mathcal{G}$ and $\mathcal{M}$, so that it is possible to use the classical group $\mathcal{G}$ to obtain the structure of **K**.

For the case in which both **X** and **Y** are *c*-tensors (and therefore **K** is again an *i*-tensor), we obtain the same result as above. The operations A_i of $\mathcal{H}$ surely provide the same relations of the type of Eq. (4–30) for both groups. But, on the other hand, so do the operations B_i and $\underline{B}_i$. Specifically, because **Y** is a *c*-tensor,

$$\mathbf{D}_Y(\underline{B}_i) = -\mathbf{D}_Y(B_i)$$

and similarly for $\mathbf{D}_X(\underline{B}_i)$. Thus, relation (4–30) is the same for $\underline{B}_i$ as it is for B_i, and the structure of **K** is the same for $\mathcal{M}$ as for $\mathcal{G}$. In both of these cases, in which **K** is an *i*-tensor, there is no need to introduce time reversal to obtain the correct structure of **K**.

Finally, we consider the case in which **Y** is an *i*-tensor and **X** a *c*-tensor, so that **K** is a *c*-tensor. Obviously, the operations A_i provide the same relations of the type (4–30), which **K** must obey for either group $\mathcal{G}$ or $\mathcal{M}$. However, the operations B_i in $\mathcal{G}$ provide different relations from $\underline{B}_i$ in $\mathcal{M}$. In order to see how it might still be possible to develop, from group $\mathcal{G}$, the structure of **K** that is applicable to $\mathcal{M}$, we introduce the following theorem, for which a proof is given in Appendix G. (It can also be verified by trial and error from the character tables, Table 3–2, with the aid of Table 5–1.)

Given a classical point group $\mathcal{G}$ composed as in Eq. (4), its irreps, γ, and its character table, then for every irrep γ, there exists an irrep γ^c, which we call the *complementary irrep* of γ, such that

$$\left.\begin{aligned} \chi^c(A_i) &= \chi(A_i) \\ \chi^c(B_i) &= -\chi(B_i) \end{aligned}\right\} \tag{5-6}$$

where $\chi(A_i)$ and $\chi(B_i)$ are the characters of γ for operations A_i and B_i, respectively, while $\chi^c(A_i)$ and $\chi^c(B_i)$ are the corresponding characters for the complementary irrep.

This theorem does not exclude the possibility that an irrep may be its own complement, in the case of a degenerate irrep. This possibility does, in fact, occur for any irrep for which $\chi(B_i) = 0$ for all B_i. On the other hand, for a one-dimensional irrep, the characters are never zero, so that the complementary irrep must be a different one from the original. Let us illustrate by examining the character table of the group C_{4v}, Table 3–2, letting the subgroup $\mathscr{H}$ be C_4 (consisting of E, $2C_4$ and C_2). The operations B_i are then the $2\sigma_v$ and $2\sigma_d$. Clearly, the two-dimensional E irrep is its own complement, while for A_1, A_2, B_1 and B_2, the complementary irreps are: A_2, A_1, B_2 and B_1 respectively.

In order to utilize this theorem, for the above-mentioned case in which **Y** is an *i*-tensor and **X** a *c*-tensor, we begin by noting that **X**, regarded as a hypervector, can be the basis of a representation Γ_X which can be decomposed into irreps:

$$\Gamma_X = \Sigma_k \gamma_k$$

(This summation may include repeated irreps, that is, a given irrep appearing more than once.) If we wish to imitate in group $\mathscr{G}$ how **X** actually transforms in group $\mathscr{M}$, it is clear that **X** must transform in $\mathscr{G}$ as the set of complementary irreps $\Gamma_X^c = \Sigma_k \gamma_k^c$, rather than as Γ_X. It then follows that the structure of the matter tensor **K** in group $\mathscr{G}$, when the complementary irreps are employed for **X**, will be the same as the structure of **K** in group $\mathscr{M}$. Of course, for the *i*-tensor **Y** the irreps of Γ_Y will be applicable. The non-zero values of **K** can then be obtained by following the methods of Chapter 4, but utilizing Γ_Y and Γ_X^c instead of Γ_Y and Γ_X.

The reverse case, where **Y** is the *c*-tensor and **X** the *i*-tensor, is equally well handled by changing the representation structure of **Y** to obtain its complementary irreps, from Γ_Y to Γ_Y^c while maintaining Γ_X.

In summary, when **K** is a *c*-tensor because *either* **X** *or* **Y** is a *c*-tensor, the structure of **K** can be obtained by working with group $\mathscr{G}$ and employing the methods of Chapter 4, but using the complementary irreps of **X** or of **Y** (whichever is the *c*-tensor).

Since in almost all cases in which this method will be applied, the *c*-tensor (**X** or **Y**) will be either the magnetic induction **B** or magnetic field, **H**, both of which are axial vectors, it is useful to know the complementary irreps of an axial vector for all magnetic groups. For this reason, we have listed these complementary irreps in the last column of Table 5–1. For the 32 singly colored groups (type I), the irreps listed are the same as those listed for the same groups in the S-C-T tables (Appendix E), but for the 58 black-and-white groups there are corresponding changes. These are obtained essentially by inspection from the character tables of the classical point groups.

We illustrate for the classical group D_{3d} whose character table is given in Table 3–2. From Table 5–1 we see that to this classical group there corresponds three variant black-and-white groups. Now, the irrep structure of an axial vector in D_{3d} is $A_{2g} + E_g$. For the variant D_{3d}: S_6 where $\{B_i\} = 3C_2$ and $3\sigma_v$, we readily find, from Table 3–2, that the complementary irrep of A_{2g} is A_{1g} while that of E_g is itself. For D_{3d}: C_{3v}, where $\{B_i\} = 3C_2$, i, and $2S_6$, $A_{2g} \to A_{1u}$ and $E_g \to E_u$ (the arrow being used to designate the complement). Finally, for D_{3d}: D_3 where $\{B_i\} =$ i, $2S_6$, and $3\sigma_v$, $A_{2g} \to A_{2u}$ and $E_g \to E_u$. In this way the reader can see how easy it is to generate the irreps in the last column of Table 5–1 by inspection of the character tables of the classical groups.

In addition to the listing of the complementary irreps of axial vector components in Table 5–1, it is also necessary to obtain their proper orientations in the case of two- and three-dimensional irreps. This has been done for all E irreps by selecting the orientation of the pair R_x, R_y to be consistent with any pair (x, y) or any pair of α_i components belonging to that same irrep already listed with its orientation in the S-C-T table (Appendix E). In so doing, we must realize that time reversal does not affect the x, y, z or α_i components, so that their orientations remain unchanged. The proper orientation of the pair R_x, R_y is shown for all E irreps by writing the pair in parentheses following the irreps in Table 5–1.

We illustrate the procedure by considering the three magnetic groups based on C_{4v} (for which the character table was given in Table 3–2). For the original group $\mathcal{G}$ we have already shown in Section 3–5 why $(R_y, -R_x)$ is the proper orientation to match (x, y) (or $(R_x, -R_y)$ to match (y, x), as given in the S-C-T Table of Appendix E). Specifically, we use C_4 and σ_v^{yz} as the two generators of the group. Now, for the first magnetic variant C_{4v}: C_4, where σ_v reverses sign, instead of $R_x \to R_x$ and $R_y \to -R_y$ as before (See page 61), we now have, because of the sign reversal: $R_x \to -R_x$, $R_y \to R_y$. Since this behavior exactly matches that of x, y, it is clear that (R_x, R_y) is similarly oriented to (x, y) in the present case (or (R_y, R_x) to (y, x)). Finally, for the variant C_{4v}: C_{2v} in which operation C_4 involves a sign reversal and σ_v does not, we obtain under C_4: $R_x \to -R_y$, $R_y \to R_x$, and under σ_v: $R_x \to R_x$, $R_y \to -R_y$. Both can be satisfied only if (R_y, R_x) is oriented similarly to (x, y) (or (R_x, R_y) to (y, x)). Examples of other groups are given in the problems.

For the cubic groups, where R_x, R_y and R_z are always partners in a three-dimensional (T-type) irrep, is easy to show that, in spite of sign changes in the complementary irreps, the orientation (R_x, R_y, R_z) still stands in all cases. Because of this result, it was deemed unnecessary to include these orientations in Table 5–1 for the T irreps.

Finally, in the case of the complex $\tilde{\mathrm{E}}$ irreps there is no freedom of choice, namely, the choice between $R_x \mp \mathrm{i}R_y$ or $R_x \pm \mathrm{i}R_y$ is set by the complex characters. For these irreps, in Table 5–1 we note which is the correct choice by writing either ($\mp$) or ($\pm$) following the irrep.

The procedure which we have developed in this section, employing complementary irreps, has made it possible to determine the structure of **K** for a magnetic group by working entirely with a classical group and employing the methods of Chapter 4. The advantage of this approach is that, in this way, we avoid the necessity for dealing with irreps of magnetic groups, a subject which involves the complexities of the theory of corepresentations (Cracknell, 1975). The reason for these complexities is that time inversion is not a unitary operation but is antiunitary; therefore, all complementary operations are antiunitary. E. P. Wigner first developed the theory of corepresentations to make possible the handling of groups containing antiunitary elements. From the viewpoint of this book, however, where the sole objective is to determine the structure of matter tensor **K**, we see that the procedure of the present chapter makes it possible to avoid involvement with corepresentation theory.

5–5 Conclusions

For static (equilibrium) phenomena, when the matter tensor **K** is time symmetric (an *i*-tensor) the determination of the structure of **K**, both for magnetic and non-magnetic crystals, can be carried out without introducing time reversal. This statement applies even to so 'magnetic' a property as the magnetic permeability, where both **X** and **Y** (here **H** and **B**) are *c*-tensors, so that the permeability $\boldsymbol{\mu}$ is an *i*-tensor.

On the other hand, when **K** is time antisymmetric we must consider magnetic and non-magnetic crystals separately. For non-magnetic crystals, such a matter tensor vanishes identically. For magnetic crystals, the structure of **K** can be determined by assigning to **X** or **Y** (whichever is the *c*-tensor) instead of its actual irreps, the complementary irreps, and then proceeding in the usual way, as described in Chapter 4. These static properties, for which **K** is time antisymmetric, have been designated 'special magnetic properties' in Table 1–1 of Chapter 1 and will be given separate treatment in Chapter 8. All other properties, which will be handled in Chapters 6, 7, 9 and 10, can be treated without consideration of time reversal.

Table 5.1. *The 90 magnetic point groups, including symmetry operations, irreps of an axial vector and orientations of* R_x, R_y *where appropriate*

Group symbol	Symmetry operations	Irreps of R_z, R_x, R_y
C_1	E	3A
C_i	E i	$3A_g$
C_i: C_1	+ −	$3A_u$
C_s	E σ_h	A′ + 2A″
C_s: C_1	+ −	A″ + 2A′
C_2	E C_2	A + 2B
C_2: C_1	+ −	B + 2A
C_{2h}	E C_2 i σ_h	$A_g + 2B_g$
C_{2h}: C_s	+ − − +	$B_u + 2A_u$
C_{2h}: C_2	+ + − −	$A_u + 2B_u$
C_{2h}: C_i	+ − + −	$B_g + 2A_g$
C_{2v}	E C_2 σ_v σ_v'	$A_2 + B_2 + B_1$
C_{2v}: C_s	+ − + −	$B_2 + A_2 + A_1$
C_{2v}: C_2	+ + − −	$A_1 + B_1 + B_2$
D_2	E $C_2(z)C_2(y)C_2(x)$	$B_1 + B_3 + B_2$
D_2: C_2	+ + − −	$A + B_2 + B_3$
D_{2h}	E $C_2(z)$ $C_2(y)$ $C_2(x)$ i σ^{xy} σ^{xz} σ^{yz}	$B_{1g} + B_{3g} + B_{2g}$
D_{2h}: C_{2v}	+ + − − − − + +	$A_u + B_{2u} + B_{3u}$
D_{2h}: D_2	+ + + + − − − −	$B_{1u} + B_{3u} + B_{2u}$
D_{2h}: C_{2h}	+ + − − + + − −	$A_g + B_{2g} + B_{3g}$
C_4	E C_4 C_2 C_4^3	A + $\tilde{E}$ (∓)
C_4: C_2	+ − + −	B + $\tilde{E}$ (±)
S_4	E S_4 C_2 S_4^3	A + $\tilde{E}$ (±)
S_4: C_2	+ − + −	B + $\tilde{E}$ (∓)
C_{4h}	E C_4 C_2 C_4^3 i S_4^3 σ_h S_4	$A_g + \tilde{E}_g$ (∓)
C_{4h}: S_4	+ − + − − + − +	$B_u + \tilde{E}_u$ (±)
C_{4h}: C_4	+ + + + − − − −	$A_u + \tilde{E}_u$ (∓)
C_{4h}: C_{2h}	+ − + − + − + −	$B_g + \tilde{E}_g$ (±)
C_{4v}	E $2C_4$ C_2 $2\sigma_v$ $2\sigma_d$	A_2 + E (R_x, $-R_y$)
C_{4v}: C_4	+ + + − −	A_1 + E (R_y, R_x)
C_{4v}: C_{2v}	+ − + + −	B_2 + E (R_x, R_y)
D_{2d}	E $2S_4$ C_2 $2C_2'$ $2\sigma_d$	A_2 + E (R_x, $-R_y$)
D_{2d}: D_2	+ − + + −	B_2 + E (R_x, R_y)
D_{2d}: C_{2v}	+ − + − +	B_1 + E (R_y, $-R_x$)
D_{2d}: S_4	+ + + − −	A_1 + E (R_y, R_x)
D_4	E $2C_4$ C_2 $2C_2'$ $2C_2''$	A_2 + E (R_x, $-R_y$)
D_4: C_4	+ + + − −	A_1 + E (R_y, R_x)
D_4: D_2	+ − + + −	B_2 + E (R_x, R_y)

Table 5.1. *(cont.)*

Group symbol	Symmetry operations	Irreps of R_z, R_x, R_y
D_{4h}	$E\ 2C_4\ C_2\ 2C_2'\ 2C_2''\ i\ 2S_4\ \sigma_h\ 2\sigma_v\ 2\sigma_d$	$A_{2g} + E_g\ (R_x, -R_y)$
$D_{4h}: D_{2d}$	+ − + + − − + − − +	$B_{2u} + E_u\ (R_x, -R_y)$
$D_{4h}: C_{4v}$	+ + + − − − − − + +	$A_{1u} + E_u\ (R_y, -R_x)$
$D_{4h}: C_{4h}$	+ + + − − + + + − −	$A_{1g} + E_g\ (R_y, R_x)$
$D_{4h}: D_4$	+ + + + + − − − − −	$A_{2u} + E_u\ (R_x, R_y)$
$D_{4h}: D_{2h}$	+ − + + − + − + + −	$B_{2g} + E_g\ (R_x, R_y)$
C_3	$E\ C_3\ C_3^2$	$A + \tilde{E}\ (\mp)$
S_6	$E\ C_3\ C_3^2\ i\ S_6^5\ S_6$	$A_g + \tilde{E}_g\ (\mp)$
$S_6: C_3$	+ + + − − −	$A_u + \tilde{E}_u\ (\mp)$
C_{3v}	$E\ 2C_3\ 3\sigma_v$	$A_2 + E\ (R_x, -R_y)$
$C_{3v}: C_3$	+ + −	$A_1 + E\ (R_y, R_x)$
D_3	$E\ 2C_3\ 3C_2$	$A_2 + E\ (R_x, -R_y)$
$D_3: C_3$	+ + −	$A_1 + E\ (R_y, R_x)$
D_{3d}	$E\ 2C_3\ 3C_2\ i\ 2S_6\ 3\sigma_d$	$A_{2g} + E_g\ (R_x, -R_y)$
$D_{3d}: S_6$	+ + − + + −	$A_{1g} + E_g\ (R_y, R_x)$
$D_{3d}: C_{3v}$	+ + − − − +	$A_{1u} + E_u\ (R_y, -R_x)$
$D_{3d}: D_3$	+ + + − − −	$A_{2u} + E_u\ (R_x, R_y)$
C_{3h}	$E\ C_3\ C_3^2\ \sigma_h\ S_3\ S_3^5$	$A' + \tilde{E}''\ (\mp)$
$C_{3h}: C_3$	+ + + − − −	$A'' + \tilde{E}'\ (\mp)$
C_6	$E\ C_6\ C_3\ C_2\ C_3^2\ C_6^5$	$A + \tilde{E}_1\ (\mp)$
$C_6: C_3$	+ − + − + −	$B + \tilde{E}_2\ (\pm)$
C_{6h}	$E\ C_6\ C_3\ C_2\ C_3^2\ C_6^5\ i\ S_3^5\ S_6^5\ \sigma_h\ S_6\ S_3$	$A_g + \tilde{E}_{1g}\ (\mp)$
$C_{6h}: C_6$	+ + + + + + − − − − − −	$A_u + \tilde{E}_{1u}\ (\mp)$
$C_{6h}: C_{3h}$	+ − + − + − − + − + − +	$B_u + \tilde{E}_{2u}\ (\mp)$
$C_{6h}: S_6$	+ − + − + − + − + − + −	$B_g + \tilde{E}_{2g}\ (\pm)$
D_{3h}	$E\ 2C_3\ 3C_2\ \sigma_h\ 2S_3\ 3\sigma_v$	$A_2' + E''\ (R_x, -R_y)$
$D_{3h}: C_{3h}$	+ + − + + −	$A_1' + E''\ (R_y, R_x)$
$D_{3h}: C_{3v}$	+ + − − − +	$A_1'' + E'\ (R_x, -R_y)$
$D_{3h}: D_3$	+ + + − − −	$A_2'' + E'\ (R_y, R_x)$
C_{6v}	$E\ 2C_6\ 2C_3\ C_2\ 3\sigma_v\ 3\sigma_d$	$A_2 + E_1\ (R_x, -R_y)$
$C_{6v}: C_6$	+ + + + − −	$A_1 + E_1\ (R_y, R_x)$
$C_{6v}: C_{3v}$	+ − + − + −	$B_2 + E_2\ (R_x, -R_y)$
D_6	$E\ 2C_6\ 2C_3\ C_2\ 3C_2'\ 3C_2''$	$A_2 + E_1\ (R_x, -R_y)$
$D_6: C_6$	+ + + + − −	$A_1 + E_1\ (R_y, R_x)$
$D_6: D_3$	+ − + − + −	$B_2 + E_2\ (R_x, -R_y)$
D_{6h}	$E\ 2C_6\ 2C_3\ C_2\ 3C_2'\ 3C_2''\ i\ 2S_3\ 2S_6\ \sigma_h\ 3\sigma_d\ 3\sigma_v$	$A_{2g} + E_{1g}\ (R_x, -R_y)$
$D_{6h}: C_{6v}$	+ + + + − − − − − − + +	$A_{1u} + E_{1u}\ (R_y, -R_x)$
$D_{6h}: C_{6h}$	+ + + + − − + + + + − −	$A_{1g} + E_{1g}\ (R_y, R_x)$
$D_{6h}: D_6$	+ + + + + + − − − − − −	$A_{2u} + E_{1u}\ (R_x, R_y)$
$D_{6h}: D_{3h}$	+ − + − − + − + − + + −	$B_{1u} + E_{2u}\ (R_y, R_x)$
$D_{6h}: D_{3d}$	+ − + − + − + − + − + −	$B_{2g} + E_{2g}\ (R_x, -R_y)$

Table 5.1. *(cont.)*

Group symbol	Symmetry operations	Irreps of R_z, R_x, R_y
T	E $4C_3$ $4C_3^2$ $3C_2$	T
T_h	E $4C_3$ $4C_3^2$ $3C_2$ i $4S_6$ $4S_6^5$ $3\sigma_h$	T_g
T_h: T	+ + + + − − − −	T_u
T_d	E $8C_3$ $3C_2$ $6S_4$ $6\sigma_d$	T_1
T_d: T	+ + + − −	T_2
O	E $6C_4$ $3C_2$ $8C_3$ $6C_2$	T_1
O: T	+ + + − −	T_2
O_h	E $8C_3$ $6C_2$ $6C_4$ $3C_2$ i $6S_4$ $8S_6$ $3\sigma_h$ $6\sigma_d$	T_{1g}
O_R: O	+ + + + + − − − − −	T_{1u}
O_h: T_h	+ + − − + + − + + −	T_{2g}
O_h: T_d	+ + − − + − + − − +	T_{2u}

Problems

5–1. For the classical group D_{2d}, verify that there are three different subgroups of order $g/2 = 4$ (namely, D_2, C_{2v} and S_4) and, therefore, that the four magnetic groups are as given in Table 5–1. Then apply the method for obtaining the magnetic groups from the one-dimensional irreps of D_{2d}, using the character tables (Table 3–2).

5–2. Do the same to obtain the magnetic variants of group C_{4h}.

5–3. Review all the crystal property tensors, **K**, listed in Table 1–1 and show which are *c*-tensors and why.

5–4. Verify the theorem on complementary irreps for each of the magnetic groups based on D_{2h} using the character tables (Table 3–2).

5–5. Do the same for the magnetic groups based on C_6 and D_{2d}.

5–6. For each of the magnetic groups based on D_{3d}, obtain the irreps of an axial vector and find the proper orientation of the pair R_x, R_y to match the S-C-T tables (Appendix E). Compare your results with the listings in Table 5–1.

5–7. Show that, for the cubic magnetic groups, the orientation (R_x, R_y, R_z) in the appropriate T irrep is always consistent with the S-C-T tables.

5–8. For each of the magnetic groups based on C_{6h}, obtain the irreps of an axial vector, and find the proper signs ($\pm$ or $\mp$) for R_x and R_y in the $\tilde{E}$ irrep.

6

Matter tensors of rank 0, 1 and 2

In this chapter, we begin to examine the effect of crystal symmetry on the forms of matter tensors of different rank, using the theory that we have developed in Chapters 3–5.

6–1 Scalar quantities (rank 0)

A matter tensor is a scalar if it relates a scalar force to a scalar response. The only example of this kind in Table 1–1 is the specific heat, C.

The treatment of such a matter tensor is trivial from the viewpoint of this book, since it provides just a single constant for every crystal, regardless of the crystal class and orientation.

6–2 Polar vector quantities (rank 1)

6–2–1 Form of the K *tensor*

In this case, we deal with a matter tensor that couples a scalar force to a vector response, or vice versa. To be specific, in the relation

$$\mathbf{Y} = \mathbf{K}\mathbf{X} \tag{6–1}$$

let us assume that **X** is the scalar quantity and **Y** the vector, as, for example, in the pyroelectric effect. In general, then, **K** is a vector and has three components. But **X** is invariant, that is, it belongs to the totally symmetric irrep A_1, while **Y** transforms as the rep Γ_{xyz}. (Here both **X** and **Y** are already symmetry coordinates.) By the Fundamental Theorem, therefore, the only components of **K** that are non-zero are those that relate **X** to components of **Y** belonging to irrep A_1. In particular, if A_1 does not appear in the decomposition of Γ_{xyz}, then **K** must be identically zero. Examination of the S-C-T tables (Appendix E) immediately shows that for all cubic classes $\mathbf{K} \equiv 0$. The same is also true for groups D_n, C_{nh}, D_{nd}, D_{nh}, S_n as well as C_i.

For groups C_n and C_{nv}, the z-component of a vector belongs to irrep A_1. Accordingly, for these groups $\mathbf{K}$ is given by

$$\mathbf{K} = \begin{pmatrix} 0 \\ 0 \\ \mathbf{K}_3 \end{pmatrix} \tag{6–2}$$

Finally, for the group C_s, both x and y belong to the totally symmetric irrep, so that

$$\mathbf{K} = \begin{pmatrix} K_1 \\ K_2 \\ 0 \end{pmatrix} \tag{6–3}$$

Of course, for the group C_1, with no symmetry operations, all three components of $\mathbf{K}$ appear.

It remains to consider, for the classes in which $\mathbf{K}$ is not identically zero, how the components of $\mathbf{K}$ depend on crystal orientation, since crystals are often grown with an arbitrary orientation. Since $\mathbf{K}$ transforms as a vector, any rotation into a new coordinate system for which the direction cosines with respect to the original system are a_{ij}, results in a new vector $\mathbf{K}'$ whose components are given by

$$\mathrm{K}_i' = a_{ij} K_j \tag{6–4}$$

Thus, for example, for groups C_n and C_{nv} for which Eq. (2) applies, we readily obtain

$$\left.\begin{aligned} K_1' &= a_{13} K_3 \\ K_2' &= a_{23} K_3 \\ K_3' &= a_{33} K_3 \end{aligned}\right\} \tag{6–5}$$

where the three components of $\mathbf{K}'$ are expressed in terms of the single constant K_3.

6–2–2 Application to pyroelectric effect; ferroelectrics

The reader should recall (from Eq. (1–10)) that the pyroelectric effect constitutes a change in electric displacement, $\mathbf{D}$, or, more conveniently, of electric polarization, $\mathbf{P}$, with a change in temperature. Pyroelectric crystals are polar crystals; they must develop polarization spontaneously by forming aligned permanent dipoles in the structure. Since the magnitude of this polarization inevitably changes with temperature, a pyroelectric effect is obtained. A much studied crystal of this category is the mineral tourmaline which belongs to the group C_{3v} and, therefore, possesses only the pyroelectric coefficient p_3 along the z-axis. The magnitude of this coefficient is 4×10^{-6} C/m^{2}K. Utilizing the appropriate dielectric susceptibility of tourmaline, it follows that the polarization produced by a ΔT of 1 K is equivalent to that produced by an electric field of 740 V/cm.

Closely related to pyroelectric crystals are the ferroelectrics which constitute a subset of the pyroelectric crystals. These are also crystals that have a spontaneous polarization, but, in addition, have bistable ionic positions so that the spontaneous polarization can be reversed by the application of a large electric field. Therefore, the P versus E curve in an ac experiment shows a hysteresis loop. Most ferroelectrics have a transition temperature, T_c, above which they lose their ferroelectric property by transformation to a higher (non-polar) symmetry group. While it was once thought that ferroelectric behavior was an anomaly, the known ferroelectric materials now number in the thousands. They are of great interest, both theoretically and for their applications, which are based on the high values of their dielectric constants, their electromechanical coupling (piezoelectric) constants and their electro-optic coefficients. We do not deal with the ferroelectric property in this book, however, because it is highly non-linear. (See Section 2–6.) For more information the reader may consult Jona and Shirane (1962) or Lines and Glass (1977).

6–3 Axial vector quantities

A matter tensor that couples a scalar force to an axial vector response, or vice versa, is itself an axial vector quantity. The only examples (see Chapter 1) are the magnetocaloric effect and its converse the pyromagnetic effect. These effects require additional consideration, however, since they are special magnetic effects. Nevertheless, for the sake of completeness we will work out the forms of the $T(1)^{ax}$ tensors. Referring to Eq. (1) and again letting $\mathbf{X}$ be the scalar quantity, we now have a response $\mathbf{Y}$ that transforms as $\Gamma_{\mathbf{R}}$, which is the rep of the three components (R_x, R_y, R_z) of an axial vector or an antisymmetric tensor of second rank (denoted Γ_{as} in Section 3–4). In general, $\mathbf{K}$ has three components, but since coupling can only take place for the irrep A_1 to which $\mathbf{X}$ belongs, we must again examine the S-C-T tables, this time for the decomposition of $\Gamma_{\mathbf{R}}$. In so doing, it immediately becomes clear that for all cubic groups $\mathbf{K} \equiv 0$. In addition, the same applies to classes C_{nv}, D_n, D_{nh} and D_{nd} for which there is no A_1 component in the decomposition of $\Gamma_{\mathbf{R}}$.

In the case of the groups C_n, C_{nh}, S_n and C_s, only R_z transforms as the A_1 irrep. Accordingly, for these classes, **K** takes the form of Eq. (2). Finally, for both triclinic groups, C_i and C_1, all three components of **K** are non-zero.

The transformation of **K** under an arbitrary change of axes is given by

$$K_i' = \pm\, a_{ij} K_j \tag{6–6}$$

where the + sign applies to transformations that leave the handedness of the axes unchanged and the − sign for those that change the handedness of the axes (from left to right or vice versa). (See Appendix A.)

6–4 Second-rank tensor quantities

A matter tensor of second rank can arise in several ways. It can result from coupling between a scalar force and a second-rank tensor response (or vice versa), as in the case of thermal expansion. Most commonly, it results from a vector–vector relationship (e.g. dielectric permeability, electrical conductivity or thermoelectric power). Finally, it can result from a relation between two axial vectors (e.g. magnetic permeability, the Hall effect). Since, from the theorem of Section 4–6, all tensors of the same rank and intrinsic symmetry have the same form for a given crystal system, we are free to explore the form of **K** as a function of crystal symmetry by using whichever relation makes the task simplest. Accordingly, in this section we will consider the vector–vector relationships, that is, the case in which both **X** and **Y** are polar vectors. In this case, both **X** and **Y** are the basis of the rep Γ_{xyz}. The transformation of these vectors to symmetry coordinates is given in the S-C-T table of Appendix E. From the Fundamental Theorem, we conclude that coefficients of $\bar{\mathbf{K}}$ only exist for irreps belonging to Γ_{xyz}. It is clear that, in this case, there are no groups for which $\mathbf{K} \equiv 0$.

6–4–1 Forms of the K *tensor for various crystal symmetries*

For all *cubic crystals*, Γ_{xyz} is one of the three-dimensional irreps labeled T, which occur only once. From Eq. (4–13), we have

$$Y_{\rm d}^{(\rm T)} = K_{\rm T} X_{\rm d}^{(\rm T)} \tag{6–7}$$

where the degeneracy index d takes on values 1, 2 and 3. Thus the same coefficient $K_{\rm T}$ relates each coordinate of **X** to the corresponding **Y** coordinate that has the same degeneracy index. For cubic crystals, therefore,

$$\mathbf{K} = \begin{pmatrix} K_T & 0 & 0 \\ 0 & K_T & 0 \\ 0 & 0 & K_T \end{pmatrix} = K_T\mathbf{I} \tag{6–8}$$

where **I** is the 3×3 identity matrix. This means that for cubic crystals, **K** is a scalar quantity, so that vector **Y** is always parallel to **X**. The result can also be expressed by stating that *any second-rank tensor property* **K** *is isotropic for cubic crystals*. It is important to note that this result applies regardless of whether **K** is or is not intrinsically symmetric. A wide variety of properties is therefore isotropic for cubic crystals, including electrical and thermal conductivity, dielectric constant, diffusivity and thermo-electric power, among others. Note that this is one of the many cases, mentioned in connection with Neumann's principle (Section 4–1), in which the crystal property exhibits higher symmetry than that of the crystal itself.

Turning now to *uniaxial crystals* (i.e. those with a single three-fold, four-fold or six-fold axis of rotation or improper rotation), we find that it is convenient to make a distinction between 'upper' and 'lower' symmetry groups. The former are those that have no irreps of type $\tilde{\mathrm{E}}$ (i.e with complex characters), while the latter do have $\tilde{\mathrm{E}}$ irreps. The following subdivision then arises:

Uniaxial groups: Upper symmetry

Hexagonal: D_{6h}, D_{3h}, D_6, C_{6v}
Tetragonal: D_{4h}, D_{2d}, D_4, C_{4v}
Trigonal: D_{3d}, D_3, C_{3v}

Uniaxial groups: Lower symmetry

Hexagonal: C_{6h}, C_{3h}, C_6
Tetragonal: C_{4h}, S_4, C_4
Trigonal: S_6, C_3

This subdivision into sets of uniaxial groups, together with the corresponding sets of the upper cubic (O_h, T_d and O) and lower cubic (T_h and T) groups, as well as the orthorhombic, the monoclinic and the triclinic sets of groups, constitute the 11 'Laue groups' that are distinguishable by X-ray diffraction. (See e.g. the International Tables for X-ray Crystallography, 1952.) The important characteristic is that the highest member of each set possesses a center of symmetry.

Examination of the S-C-T tables shows that for all the upper uniaxial groups

$$\Gamma_{xyz} = (\mathrm{A\ or\ B}) + \mathrm{E} \tag{6–9}$$

where the coordinate z belongs to the one-dimensional irrep (A or B), while x and y are partners in the two-dimensional irrep (E). Then, by the Fundamental Theorem (recalling that index 3 corresponds to z, while 1 and 2 correspond to x and y, respectively),

$$Y_3 = K_{\mathrm{A/B}} X_3$$
$$Y_1 = K_{\mathrm{E}} X_1$$
$$Y_2 = K_{\mathrm{E}} X_2$$

or

$$\mathbf{K} = \overline{\mathbf{K}} = \begin{pmatrix} K_{\mathrm{E}} & 0 & 0 \\ 0 & K_{\mathrm{E}} & 0 \\ 0 & 0 & K_{\mathrm{A/B}} \end{pmatrix} \tag{6–10}$$

This form of **K** with two independent constants, applies to all of the upper uniaxial groups. It shows that **K** is diagonal, but not a scalar quantity; the property K is, therefore, in general, a function of crystal orientation. It is, however, isotropic in the basal plane, since any rotation about the z-axis leaves the form of **K** unchanged. (See Problem 6–2.)

It is often customary to refer to $K_{\mathrm{A/B}}$ as $K_{\parallel}$, that is, the property K measured parallel to the major symmetry axis of the crystal. Similarly, K_{E} is referred to as $K_{\perp}$, symbolizing the property in the plane perpendicular to the major symmetry axis. These two quantities can be very different, in which case the crystal property K becomes highly anisotropic. (See Section 6–4–4.)

Again, it is noteworthy that for these upper uniaxial symmetries, **K** turns out to be symmetric regardless of whether or not K_{ij} is intrinsically symmetric.

Next, we consider the lower uniaxial crystals. In each of these cases

$$\Gamma_{xyz} = (\mathrm{A\ or\ B}) + \tilde{\mathrm{E}} \tag{6–11}$$

that is, z belongs to a one-dimensional irrep as before, but now x and y belong to the complex pair of one-dimensional irreps denoted by $\tilde{\mathrm{E}}$. In general, from Eqs. (4–22, 4–23) we obtain

$$\overline{K} = \begin{pmatrix} K_{\tilde{\mathrm{E}}}^{\mathrm{re}} & K_{\tilde{\mathrm{E}}}^{\mathrm{im}} & 0 \\ -K_{\tilde{\mathrm{E}}}^{\mathrm{im}} & K_{\tilde{\mathrm{E}}}^{\mathrm{re}} & 0 \\ 0 & 0 & K_{\mathrm{A/B}} \end{pmatrix} \tag{6–12}$$

where $K_{\tilde{\mathrm{E}}}^{\mathrm{re}}$ and $K_{\tilde{\mathrm{E}}}^{\mathrm{im}}$ are the real and imaginary parts of the complex quantity $K_{\tilde{\mathrm{E}}}$. In this case, it becomes necessary to distinguish whether or not **K** is intrinsically symmetric. As already mentioned in Chapter 4, if **K** is intrinsically symmetric $K_{\tilde{\mathrm{E}}}^{\mathrm{im}} = 0$, and the **K** tensor then takes on the same diagonal form (Eq. (10)) as for the upper uniaxial crystals, with just two independent constants, that is, if **K** is a $T_S(2)$ tensor, there is no distinction between the upper and lower uniaxial classes. If, on the other hand, **K** is not symmetric, then $K_E^{im} \neq 0$ and the lower uniaxial crystals have three independent constants instead of two.

We next consider *orthorhombic crystals*: D_{2h}, D_2 and C_{2v}. In each of these cases x, y and z belong to a different one-dimensional irrep, or

$$\Gamma_{xyz} = \gamma_1 + \gamma_2 + \gamma_3 \tag{6–13}$$

In this case, **K** is diagonal with three independent constants, and can be expressed in the form

$$\mathbf{K} = \begin{pmatrix} K_{\gamma_2} & 0 & 0 \\ 0 & K_{\gamma_3} & 0 \\ 0 & 0 & K_{\gamma_1} \end{pmatrix} \tag{6–14}$$

(Note that, following the convention used for the uniaxial crystals, we have taken γ_1 as the irrep of z.) For the three orthorhombic crystals classes the one-dimensional irreps are given by

	γ_1	γ_2	γ_3
D_{2h}	B_{1u}	B_{3u}	B_{2u}
D_2	B_1	B_3	B_2
C_{2v}	A_1	B_1	B_2

For the *monoclinic crystals*, C_{2h} and C_2, z belongs to A_u or A, respectively, while x and y both belong to the same irrep, B_u or B. We are then dealing with a repeated one-dimensional irrep. Table 4–1 shows that the B irrep gives rise to three constants if **K** is symmetric and four constants if it is not. If we write

$$\Gamma_{xyz} = A + 2B \tag{6–15}$$

we obtain

$$\mathbf{K} = \begin{pmatrix} K_{B,11} & K_{B,12} & 0 \\ K_{B,21} & K_{B,22} & 0 \\ 0 & 0 & K_A \end{pmatrix} \tag{6–16}$$

with

$$K_{B,12} = K_{B,21} \quad \text{(symm.)}$$

in the case where **K** is symmetric.

For the monoclinic crystal C_s, a similar result is obtained, since now

$$\Gamma_{xyz} = A'' + 2A'$$

Table 6–1. *Forms of a $T(2)$ matter tensor for various crystal symmetries*

Triclinic $\begin{pmatrix} K_{11} & K_{12} & K_{13} \\ K_{21} & K_{22} & K_{23} \\ K_{31} & K_{32} & K_{33} \end{pmatrix}$	*Lower uniaxial* $\begin{pmatrix} K_{11} & K_{12} & 0 \\ -K_{12} & K_{11} & 0 \\ 0 & 0 & K_{33} \end{pmatrix}$
Monoclinic $\begin{pmatrix} K_{11} & K_{12} & 0 \\ K_{21} & K_{22} & 0 \\ 0 & 0 & K_{33} \end{pmatrix}$	*Upper uniaxial* $\begin{pmatrix} K_{11} & 0 & 0 \\ 0 & K_{11} & 0 \\ 0 & 0 & K_{33} \end{pmatrix}$
Orthorhombic $\begin{pmatrix} K_{11} & 0 & 0 \\ 0 & K_{22} & 0 \\ 0 & 0 & K_{33} \end{pmatrix}$	*Cubic* $\begin{pmatrix} K_{11} & 0 & 0 \\ 0 & K_{11} & 0 \\ 0 & 0 & K_{11} \end{pmatrix}$

For $T_S(2)$ tensors, $K_{ij} = K_{ji}$. In particular, lower and upper uniaxials become the same.

with z belonging to irrep A″ and x and y to A′. The form of (16) still stands with A → A″ and B → A′.

Finally, for triclinic crystals C_i or C_1, x, y and z all belong to the same one-dimensional irrep. For C_i, $\Gamma = 3A_u$ while for C_1, $\Gamma = 3A$. In such cases of a three-fold repeat of the irrep, there can be no simplification of the **K** tensor due to crystal symmetry.

The forms of **K** as a $T(2)$ tensor, for all the crystal systems in the standard K_{ij} notation, are summarized in Table 6–1.

6–4–2 Property K in an arbitrary direction

The form of **K** in the standard coordinate system gives the values of the property K in these directions. It is useful, however, to obtain the results for the tensor components when the crystal is in an arbitrary orientation. Since **K** is of second rank, its components transform according to

$$K'_{ij} = a_{ik} a_{jl} K_{kl} \tag{6–17}$$

where the unprimed components are in the standard orientation, the primed components are in the actual orientation, and the a_{ij} are direction cosines between the two sets of axes. An important case is that of uniaxial crystals where **K** is in the form of Eq. (10). (Recall that this covers all uniaxial crystals if **K** is intrinsically symmetric or the upper uniaxial crystals if it is not symmetric.) Using the terminology $K_{\parallel}$ and $K_{\perp}$, as discussed following Eq. (10), we readily obtain

$$K'_{11} = K_{\perp} \sin^2 \theta + K_{\parallel} \cos^2 \theta \qquad (6\text{–}18)$$

and

$$K'_{12} = a_{13}a_{23}(K_{\parallel} - K_{\perp}) \qquad (6\text{–}19)$$

where θ is the angle between direction 1 of the crystal coordinates and the major symmetry axis of the crystal. (Note that Eq. (19) was obtained with the aid of Eqs. (17) and (A–7).) Equation (18) is useful in giving the value of the property K along a selected axis of the crystal, and its anisotropy as this axis is varied. Equation (19), on the other hand, shows that off-diagonal components appear for such a uniaxial crystal when the crystal axes lie in arbitrary directions, and that the magnitudes of these components depend on the 'anisotropy factor' $K_{\parallel} - K_{\perp}$.

For crystals of lower symmetry, results for K'_{11} and K'_{12} can, of course, be obtained directly from Eq. (17).

6–4–3 Further remarks on T(2) matter tensors

It is shown in Appendix A that any second-rank symmetric tensor can be described in terms of a tensor ellipsoid, which has three principal axes. The results of this section show that for uniaxial and orthorhombic crystals, the principal axes of the ellipsoid coincide with the crystallographic symmetry axes. For monoclinic crystals only one principal axis lies parallel to the C_2 symmetry axis (or perpendicular to the plane of symmetry) and the other two are arbitrary, that is, they depend on the values of the K_{ij} coefficients. Finally, for triclinic crystals, all three principal axes are arbitrary.

As already mentioned, the considerations of this section apply to all second-rank tensors, regardless of whether or not they relate two polar vectors. The various matter tensors of interest include the equilibrium properties of Table 1–1 and the transport properties of Table 2–1. Also, we have been able to cover both intrinsically symmetric and non-symmetric tensors (denoted $T_S(2)$ and $T(2)$, respectively, in Tables 1–1 and 2–1) in the same treatment. Included are principal effects, such as dielectric and magnetic permeability or electrical conductivity, and cross effects, such as thermoelectric power and the Hall effect. These diverse properties have in common only that they are described in terms of a second-rank tensor. Nevertheless, they are mathematically similar, with respect to the role of crystal symmetry.

It should be noted that, since the relation between two axial vectors is a second-rank tensor, the results for the form of the **K** tensor, for different crystal groups, apply as well to a property such as the magnetic permeability. The irrep labels of the K_γ constants may be different, however, since X and Y are now each the basis for the rep $\Gamma_{\mathbf{R}}$ rather than Γ_{xyz}. In the case of a property, such as thermal expansion, that couples a scalar and a second-rank tensor, again the standard forms of the **K** tensor will be those given in Table 6–1, but the irrep designations of K_γ used in deriving them will not be applicable. The useful final results, however, are those in Table 6–1, and the specific irreps involved in the derivation of these results are not important in the final analysis.

6–4–4 Application to diffusivity and electrical conductivity

We focus here primarily on uniaxial crystals, for which there are two constants, $K_\parallel$ and $K_\perp$. The degree of anisotropy of the property is then based on how much these constants differ. All that our theory, based on symmetry, can tell us is that these two constants cannot be exactly the same, that is, that the property is not isotropic, as it is for cubic crystals. But the extent of anisotropy depends on atomic structure and the detailed mechanisms of the phenomena, topics that are outside the scope of this book. Nevertheless, in order to give the reader some perspectives, we will present some pertinent examples.

For non-cubic metals, the anisotropy of electrical resistivity is usually not large. For the case of hexagonal zinc, for example, the resistivity perpendicular to the c-axis is only 3% lower than that parallel to this axis.

Other uniaxial materials, particularly ceramic crystals, can show high degrees of anisotropy of the resistivity. Some materials have a structure such that rapid motion of carriers is possible within planes perpendicular to the c-axis, but with difficulty for these carriers to cross those planes. Such materials constitute *two-dimensional* conductors or diffusers. A notable example is β-alumina, whose formula is close to $MAl_{11}O_{17}$, where M is a monovalent ion, most often an alkali ion. The structure consists of several layers of Al_2O_3 with the gamma alumina (spinel) structure connected by a bridging layer containing only the M^+ and O^{2-} ions. The M^+ ions migrate easily in the bridging layers, making these materials 'fast-ion conductors', but move out of the bridging planes with great difficulty. Accordingly, the electrical conductivity is highly anisotropic with $\sigma_\perp \gg \sigma_\parallel$.

Another interesting and important example is the most widely studied high-temperature superconductor $YBa_2Cu_3O_{7-x}$ (or YBCO), whose superconducting transition temperature, T_c, is near 90 K. The structure is dominated by Cu–O-planes which are perpendicular to the *c*-axis. Actually, the structure is orthorhombic, so there are three coefficients (see Table 6–1). We look at the normal electrical resistivity (above T_c), as shown in Fig. 6–1. Here, no distinction is made between the *a* and *b* directions, because there is not a large difference, and because crystals tend to be twinned rather than perfect. The diagram shows that the *c*-axis resistivity (ρ_c) is about 30 times larger than that of *a* and *b* (ρ_{ab}) near room temperature. But it also shows that the temperature coefficients of these two properties are very different, ρ_{ab} showing the usual linear dependence on temperature representing metallic behavior, while ρ_c has a very small temperature coefficient (and, in fact, a negative one just above T_c). This result emphasizes that the two resistivities are truly independent properties, as our symmetry analysis has shown.

Oxygen diffusion in YBCO is even more striking, the work of Rothman *et al.* (1991) showing that ^{18}O tracer diffusion gives $D_c \ll D_{ab}$ with anisotropy ratio as high as 10^4–10^6.

By contrast with the two-dimensional conductors and diffusers, a number of uniaxial materials show strong anisotropy in the reverse direction with $K_{\parallel} \gg K_{\perp}$. This is usually the result of the existence of open channels

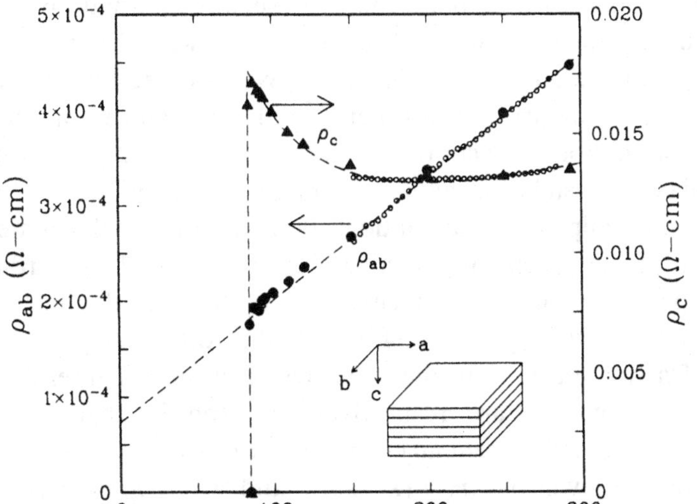

Fig. 6–1. Resistivity components parallel to Cu–O planes (ρ_{ab}) and perpendicular to them (ρ_c) for $YBa_2Cu_3O_{7-x}$ crystals above the superconducting transition temperature. Note the two different scales of resistivity. (After Tozer *et al.*, 1987).

along the z-axis. Such materials are *one-dimensional* conductors or diffusers. A notable example comes from a study of Li^+ diffusion in tetragonal rutile (TiO_2) crystals (Johnson, 1964). Here, because of the existence of channels in the structure, $D_{\parallel}$ is high, having a low activation energy of 0.33 eV, while $D_{\perp}$ is at least 10^8 times smaller (and barely measurable).

Other examples of one-dimensional ionic migration occur in tetragonal and hexagonal tungsten bronzes, which have the general formula M_xWO_3, where M is usually a monovalent cation and $0 < x < 1$. These materials are mixed conductors, both ionic and electronic, and, as such, are useful as reversible electrodes in systems containing a purely ionic conductor as the electrolyte (e.g. β-alumina). The ionic component of the conductivity, however, has been shown to involve migration down channels that are parallel to the c-axis, so that $\sigma_{\parallel} \gg \sigma_{\perp}$. (See Hagenmuller and van Gool, 1978.)

6–4–5 Application to the optical indicatrix

In the case of optical properties, the ellipsoid that we deal with is the indicatrix, which is based on the impermeability tensor, $\boldsymbol{\beta}$ (see Section 1–6). It should be recalled that double refraction occurs when the central section of the indicatrix perpendicular to the wave-normal direction is an ellipse. Further, the refractive indices of the two wave fronts are given by the semiaxes of the ellipse (see Fig. 1–1). We now see that in the case of cubic crystals, the indicatrix is a sphere; in this case all central sections are circles, so that double refraction does not occur. A more interesting case is that of uniaxial crystals (hexagonal, tetragonal or trigonal) for which the indicatrix is an ellipsoid of revolution about the principal symmetry axis, that is, two of the principal refractive indices are equal. Accordingly, for waves propagated along the principal axis there is no double refraction; this axis is called the *optic axis*. For all other directions of propagation double refraction occurs. If the dielectric constants perpendicular and parallel to the optic axis are, respectively, $K_{\perp}$ and $K_{\parallel}$, then the refractive indices of the so-called *ordinary* and *extraordinary* waves, n_o and n_e, respectively, are given by: $n_o = (\kappa_{\perp}/\kappa_0)^{1/2}$ and $n_e = (\kappa_{\parallel}/\kappa_0)^{1/2}$ (where κ_0 is the vacuum permittivity). If $n_e > n_o$, the crystal is said to be *positive*, and if $n_e < n_o$, it is *negative*. The wave surface of the ordinary wave is a sphere, while that of the extraordinary wave is an ellipsoid of revolution that lies inside the sphere and touches it in the direction of the optic axis. This is because the refractive index of the extraordinary wave varies from

Table 6–2. *Refractive indices n_o and n_e of some uniaxial crystals (after Shuvalov, 1988)*

Crystal	n_o	n_e
Ice, H_2O	1.309	1.313
Quartz, SiO_2	1.544	1.553
Calcite, $CaCO_3$	1.658	1.486
Corundum, Al_2O_3	1.768	1.760
Rutile, TiO_2	2.616	2.903
$LiNbO_3$	2.272	2.187

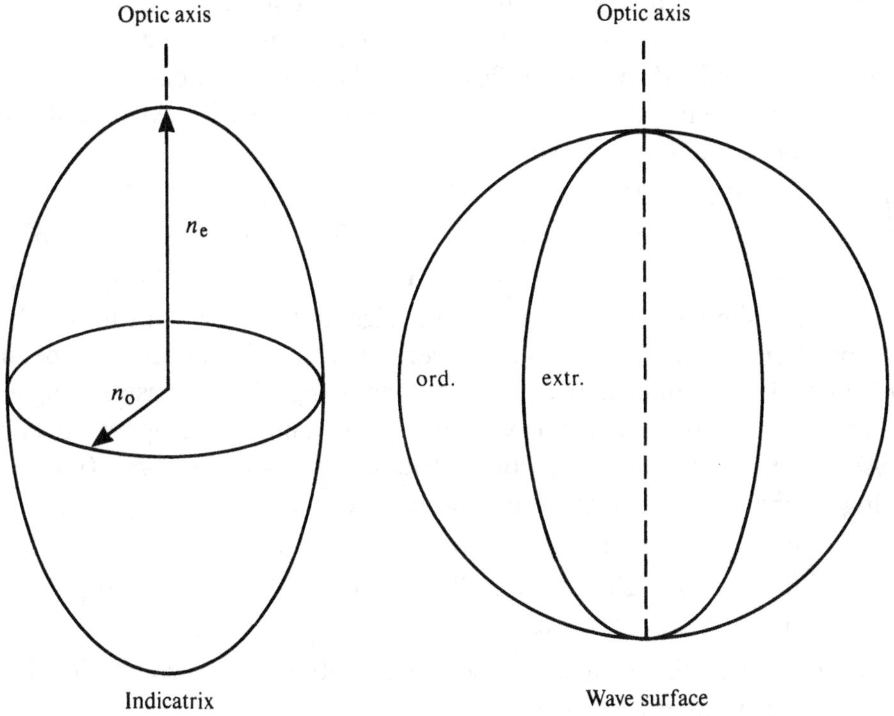

Fig. 6–2. The indicatrix and wave surface for a positive uniaxial crystal.

n_o to n_e according to the direction of the wave normal. Figure 6–2 shows the indicatrix and the wave surface for a positive uniaxial crystal.

Some numerical values for n_o and n_e of a number of uniaxial crystals are given in Table 6–2. It should be noted that both positive and negative crystals are represented in this table.

For all other crystal systems of lower symmetry (orthorhombic, monoclinic and triclinic), the indicatrix is a triaxial ellipsoid. For such crystals, it follows that there are two directions of the ellipsoid for which the central sections are circles and, therefore, along which there is no double refraction. Because of the presence of two optic axes, such crystals are said to be *biaxial*. However, it is important to note that the direction of the two optic axes is not related to the symmetry axes of the crystal.

6–4–6 Application to the Hall tensor (see Section 2–4)

We have regarded the Hall tensor in terms of two-index notation, R_{ik}, so that it constitutes a $T(2)$ tensor that is not symmetric (nine components in the triclinic case). However, in the case of the Hall tensor (as well as the analogous Righi–Leduc tensor) it is customary to return to the three-index form, R_{ijk} as defined by Eq. (2–26). Accordingly, the scheme for R_{ijk} for the triclinic system is a set of three antisymmetric 3×3 blocks placed side by side, for $k = 1$, 2 and 3 respectively, as follows:

$$\begin{pmatrix} 0 & 121 & 131 & 0 & 122 & 132 & 0 & 123 & 133 \\ -121 & 0 & 231 & -122 & 0 & 232 & -123 & 0 & 233 \\ -131 & -231 & 0 & -132 & -232 & 0 & -133 & -233 & 0 \end{pmatrix}$$

Here, for simplicity, we show only the sign and the subscript ijk of R_{ijk}, except where that coefficient is equal zero. This matrix has nine independent coefficients obtained from the two-index R_{ik}. In the case of higher symmetries the above matrix is simplified; for example, for upper uniaxial crystals, where the two-index coefficients are $R_{11} = R_{22}$ and R_{33} (the rest being = 0), we obtain for the three-index array

$$\begin{pmatrix} 0 & 0 & 0 & 0 & 0 & -231 & 0 & 123 & 0 \\ 0 & 0 & 231 & 0 & 0 & 0 & -123 & 0 & 0 \\ 0 & -231 & 0 & 231 & 0 & 0 & 0 & 0 & 0 \end{pmatrix}$$

For this case, the Hall effect equations, obtained by combining Eq. (2–3) with Eq. (2–26), become

$$\left.\begin{aligned} E_1 &= R_{123} j_2 B_3 - R_{231} j_3 B_2 \\ E_2 &= R_{231} j_3 B_1 - R_{123} j_1 B_3 \\ E_3 &= R_{231}(j_1 B_2 - j_2 B_1) \end{aligned}\right\} \qquad (6\text{–}20)$$

Finally, for the case of cubic (or isotropic) materials, only one coefficient remains, namely, $R_{123} = R_{231}$.

It is worth pointing out that the Hall coefficients, unlike most of the other crystal properties that we deal with, are highly 'defect sensitive', especially in the case of semiconductors. That is, most properties that we have been considering are defect insensitive, reflecting the symmetry, atomic arrangement and the nature of bonding in the perfect crystal and are only modified in their second, or even third, significant figures by the presence of defects. Defect-sensitive properties, on the other hand, depend primarily on departures from the perfect crystal. In the case of n-type semiconductors, which conduct by electrons introduced by donor dopants, the Hall coefficient R_H is given by $1/ne$, where n is the electron (or hole) concentration and e the electronic charge. This coefficient is then very dopant dependent. When the Hall effect is combined with the conductivity (which depends on the product of n and the carrier mobility), it becomes possible to obtain the mobility.

6–5 Second-rank axial tensors

The remaining task of this chapter is to consider matter tensors that couple a polar vector to an axial vector. Such properties are denoted by $T(2)^{ax}$ in Table 1–1, the principal example being optical activity.* (Magnetoelectric polarizability also falls into this category, but it cannot be considered here since it is a special magnetic property.)

6–5–1 Forms of the **K** *tensor for various crystal symmetries*

To be specific, let us choose **X** as the polar vector and **Y** as the axial vector. Thus **X** is the basis for rep Γ_{xyz}, while **Y** is the basis for $\Gamma_{\mathbf{R}}$. It is then clear that **K** can have non-zero components only for irreps that are shared in common between these two reps. Any crystal class for which Γ_{xyz} and $\Gamma_{\mathbf{R}}$ do not share at least one common irrep must have $\mathbf{K} \equiv 0$. For example, for any crystal that has a center of symmetry, Γ_{xyz} always belongs to the irreps of odd parity (labeled sub u) while $\Gamma_{\mathbf{R}}$ belongs to the irreps of even parity (labeled sub g). Such crystals, therefore, cannot display the property K. In other words, the presence of a center of inversion symmetry is a sufficient condition to give $\mathbf{K} \equiv 0$. But there are other possibilities for meeting the condition for $\mathbf{K} \equiv 0$. In fact, of the 32 crystal point groups only 15 are capable of exhibiting optical activity.

* Optical activity is, in fact, a symmetric tensor $T_s(2)^{ax}$ since it relates an axial scalar, $T(0)^{ax}$, to a second-rank symmetric tensor, $T_s(2)$. As a matter of convenience, we choose to derive the forms of **K** from the $T(1)$–$T(1)^{ax}$ relationship since this also gives a $T(2)^{ax}$ matter tensor and is easier to work with. It should be recognized, however, that with an actual $T(1)$–$T(1)^{ax}$ relationship, the resulting $T(2)^{ax}$ matter tensor cannot be symmetric.

Starting from the cubic groups, we note from the S-C-T tables that only for groups O and T is there a non-zero **K** tensor, and for these two groups Γ_{xyz} and $\Gamma_{\mathbf{R}}$ belong to the same three-dimensional irrep. Thus, for these two groups, the form of **K** is the same as that of Eq. (8), and **K** behaves as a scalar quantity.

Among the uniaxial groups, several have $\mathbf{K} \equiv 0$ for the reasons stated above. These include D_{nh}, C_{nh} (for $n = 3$, 4 and 6) and also D_{3d} and S_6. For the groups D_n, corresponding components of a polar vector (x, y, z) and an axial vector (R_x, R_y, R_z) transform as the same irreps. (This occurs because the D_n groups have only rotations, under which polar and axial vectors transform in the same manner.) Accordingly, for the D_n groups, the form of **K** is the same as that of a second-rank tensor, that is, as given by Eq. (10). The same may be said for the groups C_n, which contain the complex pairs of one-dimensional irreps denoted by $\tilde{E}$. All these cases are listed in Table 6–3.

For the uniaxial groups C_{nv}, we note that z and R_z belong to different irreps, so that $K_{33} = 0$. But the pair (x, y) is matched by $(R_y, -R_x)$ both belonging to the same degenerate E-type irrep. It follows immediately, from the Fundamental Theorem, Section 4–4, that

$$\left.\begin{aligned} Y_2^{(E)} &= K_E X_1^{(E)} \\ Y_1^{(E)} &= -K_E X_2^{(E)} \end{aligned}\right\} \qquad (6\text{–}21)$$

which means that $\bar{\mathbf{K}}$ takes the form

$$\bar{\mathbf{K}} = \mathbf{K} = \begin{pmatrix} 0 & -K_E & 0 \\ K_E & 0 & 0 \\ 0 & 0 & 0 \end{pmatrix} \qquad (6\text{–}22)$$

If we impose the additional requirement that K be intrinsically symmetric, as in the case of optical activity, it follows that $\mathbf{K} \equiv 0$.

Turning now to the tetragonal group D_{2d}, for which, again, $K_{33} = 0$ but now (x, y) is matched by $(R_x, -R_y)$, we find that $\bar{\mathbf{K}}$ takes the form

$$\bar{\mathbf{K}} = \mathbf{K} = \begin{pmatrix} K_E & 0 & 0 \\ 0 & -K_E & 0 \\ 0 & 0 & 0 \end{pmatrix} \qquad (6\text{–}23)$$

This result is equally valid whether or not **K** possesses intrinsic symmetry. These **K** tensors, in standard K_{ij} notation, are listed in Table 6–3. These two examples are good ones to illustrate the importance of having the correct orientation of the symmetry coordinates belonging to a two-dimensional E irrep. Note how the forms of Eqs. (22) and (23) depend critically on how these orientations are matched.

For the last of the uniaxial crystals, consider the group S_4. In this case X_1 and X_2 belong to the complex pair $\tilde{\mathrm{E}}$ with symmetry coordinates $X_1 - \mathrm{i}X_2$ and $X_1 + \mathrm{i}X_2$, while the corresponding Y coordinates are $Y_1 + \mathrm{i}Y_2$ and $Y_1 - \mathrm{i}Y_2$. This situation is then that covered by Eq. (4–25) in that, now, opposite complex quantities are matched. In the present case we therefore obtain:

$$\left.\begin{aligned} Y_1 &= K_{\tilde{\mathrm{E}}}^{\mathrm{re}} X_1 + K_{\tilde{\mathrm{E}}}^{\mathrm{im}} X_2 \\ Y_2 &= K_{\tilde{\mathrm{E}}}^{\mathrm{im}} X_1 - K_{\tilde{\mathrm{E}}}^{\mathrm{re}} X_2 \end{aligned}\right\} \tag{6–24}$$

This yields for the $\bar{\mathbf{K}}$ tensor

$$\bar{\mathbf{K}} = \begin{pmatrix} K_{\tilde{\mathrm{E}}}^{\mathrm{re}} & K_{\tilde{\mathrm{E}}}^{\mathrm{im}} & 0 \\ K_{\tilde{\mathrm{E}}}^{\mathrm{im}} & -K_{\tilde{\mathrm{E}}}^{\mathrm{re}} & 0 \\ 0 & 0 & 0 \end{pmatrix} \tag{6–25}$$

Since this result is symmetric, it applies equally well whether or not $\mathbf{K}$ is intrinsically symmetric. The result is quite different from what would be obtained from Eqs. (4–24) and serves to reemphasize the importance of correctly listing the symmetry coordinates in the S-C-T tables.

Turning now to the orthorhombic crystals, we first note that for the D_{2h} group, $\mathbf{K} \equiv 0$. The D_2 group gives the same result as for second-rank tensors, given by Eq. (14) with irreps from the table just below this equation. This $\mathbf{K}$ tensor is in diagonal and, therefore, symmetric form. For the class C_{2v}, $\bar{\mathbf{K}}$ takes the form

$$\bar{\mathbf{K}} = \mathbf{K} = \begin{pmatrix} 0 & K_{\mathrm{B}_1} & 0 \\ K_{\mathrm{B}_2} & 0 & 0 \\ 0 & 0 & 0 \end{pmatrix} \tag{6–26}$$

for the non-symmetric case. If $\mathbf{K}$ is intrinsically symmetric, we must also have

$$K_{\mathrm{B}_1} = K_{\mathrm{B}_2} \quad \text{(symm.)} \tag{6–27}$$

For monoclinic crystals, it is clear that class C_{2h} gives $\mathbf{K} \equiv 0$. In the case of group C_2, z and R_z belong to A while x, y, and R_x, R_y belong to B (a repeated irrep). The form of $\mathbf{K}$ is therefore the same as for the $T(2)$ case, Eq. (16). For the C_s group, a different result is obtained. Here x, y and R_z belong to A′ while z, R_x and R_y belong to A''. With the aid of Eqs. (4–14, 4–15) we readily obtain

$$\bar{\mathbf{K}} = \mathbf{K} = \begin{pmatrix} 0 & 0 & K_{\mathrm{A}'',1} \\ 0 & 0 & K_{\mathrm{A}'',2} \\ K_{\mathrm{A}',1} & K_{\mathrm{A}',2} & 0 \end{pmatrix} \tag{6–28}$$

Table 6–3. *Forms of a* $T(2)^{ax}$ *matter tensor for various crystal classes*

C_1 $\begin{pmatrix} K_{11} & K_{12} & K_{13} \\ K_{21} & K_{22} & K_{23} \\ K_{31} & K_{32} & K_{33} \end{pmatrix}$	D_2 $\begin{pmatrix} K_{11} & 0 & 0 \\ 0 & K_{22} & 0 \\ 0 & 0 & K_{33} \end{pmatrix}$	C_{nv} $(n = 3, 4, 6)$ $\begin{pmatrix} 0 & K_{12} & 0 \\ -K_{12} & 0 & 0 \\ 0 & 0 & 0 \end{pmatrix}$
C_s $\begin{pmatrix} 0 & 0 & K_{13} \\ 0 & 0 & K_{23} \\ K_{31} & K_{32} & 0 \end{pmatrix}$	S_4 $\begin{pmatrix} K_{11} & K_{12} & 0 \\ K_{12} & -K_{11} & 0 \\ 0 & 0 & 0 \end{pmatrix}$	D_n $(n = 3, 4, 6)$ $\begin{pmatrix} K_{11} & 0 & 0 \\ 0 & K_{11} & 0 \\ 0 & 0 & K_{33} \end{pmatrix}$
C_2 $\begin{pmatrix} K_{11} & K_{12} & 0 \\ K_{21} & K_{22} & 0 \\ 0 & 0 & K_{33} \end{pmatrix}$	D_{2d} $\begin{pmatrix} K_{11} & 0 & 0 \\ 0 & -K_{11} & 0 \\ 0 & 0 & 0 \end{pmatrix}$	O and T $\begin{pmatrix} K_{11} & 0 & 0 \\ 0 & K_{11} & 0 \\ 0 & 0 & K_{11} \end{pmatrix}$
C_{2v} $\begin{pmatrix} 0 & K_{12} & 0 \\ K_{21} & 0 & 0 \\ 0 & 0 & 0 \end{pmatrix}$	C_n $(n = 3, 4, 6)$ $\begin{pmatrix} K_{11} & K_{12} & 0 \\ -K_{12} & K_{11} & 0 \\ 0 & 0 & K_{33} \end{pmatrix}$	

For all other classes $\mathbf{K} \equiv 0$. For symmetric $T_S(2)^{ax}$, $K_{ij} = K_{ji}$. Then, for C_{nv} $(n = 3, 4, 6)$, $\mathbf{K} \equiv 0$.

and when **K** is symmetric, we also require

$$\left.\begin{aligned} K_{A'',1} &= K_{A',1} \\ K_{A'',2} &= K_{A',2} \end{aligned}\right\} \text{(symm)} \tag{6–29}$$

For triclinic crystals, the group C_i has $\mathbf{K} \equiv 0$ (as do all crystals with a center of symmetry), while for C_1 there is no simplification of the full **K** tensor.

All these $T(2)^{ax}$ tensors are listed in Table 6–3 in the standard K_{ij} notation.

Let us now consider how the form of **K** varies when the crystal is taken into an arbitrary orientation. For a second-rank axial tensor, we have (see Appendix A)

$$K'_{ij} = \pm\, a_{ik} a_{jl} K_{kl} \tag{6–30}$$

with $\pm$ signs according to whether the transformation does or does not preserve the handedness of the axes. Accordingly, in the same manner as discussed below Eq. (17) for the $T(2)$ tensors, we may obtain components K'_{ij} in arbitrary crystal directions.

6–5–2 Application to optical activity (see Section 1–7): case of quartz

For uniaxial crystals, the most conspicuous manifestation of optical activity is observed for light propagated along the optic axis. (In all other directions of light propagation, the optical activity is manifested in additional ellipticity of the transmitted light as compared with that due to the linear birefringence alone; accordingly, its observation becomes rather difficult.)

A good example of an optically active crystal is α-quartz (point group D_3). This crystal is also enantiomorphous, that is, it may occur in two forms: right- and left-handed quartz. In order to be enantiomorphous a crystal must have no element of symmetry (inversion or mirror) that changes the handedness. This is possible only for point groups of types C_n and D_n which consist entirely of rotations, and for which a psendoscalar quantity belongs to the totally symmetric (A_1-type) irrep. In addition, however, the atomic arrangement must provide a screw axis so as to produce a helical distribution of atoms. For such crystals, the optical activity of the two types of handedness will have opposite signs.

In the case of quartz, for wavelength, λ, of 510 μm, the components of the gyration tensor are $g_{11} = \pm 5.82 \times 10^{-5}$ and $g_{33} = \mp 12.96 \times 10^{-5}$, the two signs representing the right- and left-handed cases, respectively (Shuvalov, 1988, Section 7–6). For the specific rotation of the plane of polarization along the optic axis, we find: $\rho = \pi g_{33}/n_o\lambda$ (where n_o is the ordinary refractive index $\doteq 1.5$) giving a value for ρ of about 30 deg/mm.

Problems

6–1. For the case where **K** is a polar vector, obtain the form of **K** for arbitrary orientation of a crystal of point group C_s (noting that in standard axes **K** is given by Eq. (3)).

6–2. Show that for the uniaxial groups, a $T(2)$ tensor given by Eq. (10) is isotropic in the basal plane, that is, that any rotation about the z-axis leaves the form of **K** unchanged.

6–3. Obtain Eqs. (18) and (19) for the components of **K** of a $T(2)$ tensor of a uniaxial crystal in arbitrary orientation.

6–4. Obtain the matrix for the three-index Hall coefficient R_{ijk} for the case of the lower uniaxial crystals. Comparing with Eq. (6–30), what extra effects can be observed for these crystals that are not present for the upper uniaxials?

6–5. Work out the form of **K** for a $T(2)^{\text{ax}}$ tensor as given by Eq. (6–28) for a monoclinic crystal belonging to the C_s group. Note what a large number of constants have become zero as the consequences of a single mirror plane!

6–6. An alternative way to obtain **K** for a $T_S(2)^{ax}$ tensor is to find common irreps of a $T_S(2)$, that is, of Γ_α, and a pseudoscalar $T(0)^{ax}$. (Note that the latter belongs to a one-dimensional irrep having characters of +1 for all rotations and −1 for operations σ, i and S_n.) Use this method to obtain the form of **K** for groups C_{nv}, D_{2d} and S_4, and compare with the results of Table 6–3.

7

Matter tensors of rank 3

7–1 Partly symmetric tensors of rank 3

In this section we deal with a property K that couples a vector force to a second-rank symmetric tensor, $T_S(2)$ response, or vice versa. Such a matter tensor is referred to in Chapter 1 as a $T(3)$ type, but is partly symmetric (in the coefficients that come from $T_S(2)$). To be specific, let **X** be the $T_S(2)$ quantity and **Y** the polar vector. In handling this problem we make use of the Weyl normalization and take **X** as a six-vector, in the manner of obtaining α_i from α_{jk} (Eq. (4–4)). Thus, **K** in the relation

$$\mathbf{Y} = \mathbf{KX} \tag{7–1}$$

is a 3×6 matrix. As in the preceding chapter, we will arrive at the form of **K** by transforming Eq. (1) to symmetry coordinates: $\bar{\mathbf{Y}} = \bar{\mathbf{K}}\bar{\mathbf{X}}$, applying the Fundamental Theorem, and then returning to the original form of **K**. For practical purposes, however, **K** is not used in this form but must be converted either to three-index quantities (as a third-rank tensor) or to a two-index 'engineering' notation. From Eq. (4–4), the conversion to three-index quantities is

$$\left.\begin{aligned} K_{ij} &\rightarrow K_{ijj} \qquad \text{for } j = 1, 2, 3 \\ K_{i4} &\rightarrow \sqrt{2}K_{i23} \end{aligned}\right\} \tag{7–2}$$

and similarly for $i5 \rightarrow i31$, $i6 \rightarrow i12$. For conversion to the 2-index engineering notation (in the case where the quantity **X** is a system of stress and not strain) we obtain (see Chapter 1, Section 1–4)

$$\left.\begin{aligned} K_{ij}^{e} &= K_{ij} \qquad \text{for } j = 1, 2, 3 \\ K_{ij}^{e} &= \sqrt{2}K_{ij} \quad \text{for } j = 4, 5, 6 \end{aligned}\right\} \tag{7–3}$$

(If **X** is a strain, the result for $j = 4, 5, 6$ is $K_{ij}/\sqrt{2}$.)

The most important properties in this category are the piezoelectric coefficients d_{ijk}, which are usually expressed in two-index engineering form d^e_{ij}, and the electro-optic coefficients. In the latter case, the two-index property yields a 6×3 matrix which is the transposed version of the matrix that we will be discussing here.

7–1–1 Form of the **K** *tensor for various crystal symmetries*

Returning to the present problem of Eq. (1), we take **Y** to be a polar vector whose rep is Γ_{xyz}, while **X** is a $T_S(2)$ tensor with rep of Γ_α. In symmetry-coordinate form, $\bar{K}$ components only exist for irreps which Γ_α and Γ_{xyz} have in common. For a crystal possessing a center of symmetry, however, all irreps of Γ_{xyz} are of the u-type (odd parity) since a polar vector reverses sign under inversion. On the other hand, all irreps of Γ_α are of the g-type (even parity). Thus we immediately conclude that *for any group possessing a center of symmetry* $\mathbf{K} \equiv 0$. Examination of the S-C-T tables (Appendix E) shows that there is one other group for which Γ_{xyz} and Γ_α have no common irreps, namely, the cubic O group, where x, y, z belong to irrep T_1 to which no α components belong. All other point groups have some non-vanishing components of **K**. We will calculate some of these as an illustration of the method.

In the case of the two *cubic groups* T and T_d for which **K** does not vanish identically, the coordinates x, y and z and also α_4, α_5 and α_6 belong to the same three-dimensional irrep (called T and T_2, respectively). Thus, by the fundamental theorem:

$$\left.\begin{aligned} Y_1^{(T)} &= K_T X_1^{(T)} \\ Y_2^{(T)} &= K_T X_2^{(T)} \\ Y_3^{(T)} &= K_T X_3^{(T)} \end{aligned}\right\} \tag{7–4}$$

in which $X_1^{(T)}$, $X_2^{(T)}$ and $X_3^{(T)}$ are X_4, X_5 and X_6 respectively. Since the symmetry coordinates are just the original coordinates, we immediately obtain the form of **K**:

$$\mathbf{K} = \begin{pmatrix} 0 & 0 & 0 & K_{14} & 0 & 0 \\ 0 & 0 & 0 & 0 & K_{14} & 0 \\ 0 & 0 & 0 & 0 & 0 & K_{14} \end{pmatrix} \tag{7–5}$$

in which $K_{14} = K_T$.

Turning to uniaxial crystals, consider the groups D_n for which $n = 4$ or 6. Referring to the S-C-T tables, we see that coordinate z belongs to an irrep (A_2) that does not appear in Γ_α. On the other hand, the pair $(x, -y)$ belongs to an E irrep, as does the pair (α_4, α_5) with a similar orientation. Thus, **K** takes the form

$$\mathbf{K} = \begin{pmatrix} 0 & 0 & 0 & K_{14} & 0 & 0 \\ 0 & 0 & 0 & 0 & -K_{14} & 0 \\ 0 & 0 & 0 & 0 & 0 & 0 \end{pmatrix} \tag{7–6}$$

in which $K_{14} = K_{\mathrm{E}}$. For group D_3, on the other hand, Γ_α involves a repeated E irrep, with symmetry-coordinate pairs (α_4, α_5) and $(\alpha_1 - \alpha_2, \alpha_6)$ having the same orientation as $(x, -y)$. Accordingly, we take as symmetry coordinates

$$\left.\begin{aligned} \bar{\mathbf{Y}} &= \{Y_1, Y_2, Y_3\} \\ \bar{\mathbf{X}} &= \{(X_1 + X_2)/\sqrt{2}, (X_1 - X_2)/\sqrt{2}, X_3, X_4, X_5, X_6\} \end{aligned}\right\} \tag{7–7}$$

In group-theoretical notation (see the S-C-T table), $\bar{X}$ becomes

$$\bar{\mathbf{X}} = \{X_1^{(\mathrm{A}_1)}, X_{11}^{(\mathrm{E})}, X_2^{(\mathrm{A}_1)}, X_{21}^{(\mathrm{E})}, X_{22}^{(\mathrm{E})}, X_{12}^{(\mathrm{E})}\}.$$

Referring to Eq. (4–18), we see that the matrix $\bar{\mathbf{K}}$ will be

$$\bar{\mathbf{K}} = \begin{pmatrix} 0 & K_{\mathrm{E},1} & 0 & K_{\mathrm{E},2} & 0 & 0 \\ 0 & 0 & 0 & 0 & -K_{\mathrm{E},2} & -K_{\mathrm{E},1} \\ 0 & 0 & 0 & 0 & 0 & 0 \end{pmatrix} \tag{7–8}$$

from which we may return to the original coordinates to obtain

$$\mathbf{K} = \begin{pmatrix} K_{11} & -K_{11} & 0 & K_{14} & 0 & 0 \\ 0 & 0 & 0 & 0 & -K_{14} & -\sqrt{2}K_{11} \\ 0 & 0 & 0 & 0 & 0 & 0 \end{pmatrix} \tag{7–9}$$

where

$$K_{11} = K_{\mathrm{E},1}/\sqrt{2} \quad \text{and} \quad K_{14} = K_{\mathrm{E},2}$$

Note that there are only two independent constants in this **K** matrix.

The step from Eq. (8) to Eq. (9) could have been carried out in a formal way by using Eq. (4–10). However, the reader will undoubtedly see that the transformation between the original coordinates and symmetry coordinates is so simple in this case that it is easily done by inspection. Table 7–1 contains this matrix, along with those for all the other crystal symmetries, in two-index notation.

The groups C_{nv} differ from D_n in that there is now an A irrep that contains z as well as $\alpha_1 + \alpha_2$ together with α_3 (repeated). For C_{4v}, the E irrep contains (y, x) as well as (α_4, α_5) in similar orientation. On the other hand, symmetry coordinate $\alpha_1 - \alpha_2$ belongs to B_1 and α_6 to B_2 for which there are no coordinates of x, y, z. If, again, we take the symmetry coordinates as in Eq. (7), and use Eq. (4–14) for the A_1 irrep, we obtain for the $\bar{\mathbf{K}}$ matrix

$$\bar{\mathbf{K}} = \begin{pmatrix} 0 & 0 & 0 & 0 & K_{\mathrm{E}} & 0 \\ 0 & 0 & 0 & K_{\mathrm{E}} & 0 & 0 \\ K_{\mathrm{A},1} & 0 & K_{\mathrm{A},2} & 0 & 0 & 0 \end{pmatrix} \tag{7–10}$$

Table 7–1. *Forms, for various crystal symmetries, of a* **K** *tensor of rank 3 that relates a* $T(1)$ *to a* $T_S(2)$*, in two-index notation. (For the piezoelectric (engineering) coefficients* d^e_{ij}*, change all factors* $\sqrt{2}$ *to factors of 2. For the electro-optic (engineering) coefficients* r^e_{ij}*, transpose into a* 6×3 *matrix and drop all factors* $\sqrt{2}$*)*

Class C_1 (18 constants)

$$\begin{pmatrix} K_{11} & K_{12} & K_{13} & K_{14} & K_{15} & K_{16} \\ K_{21} & K_{22} & K_{23} & K_{24} & K_{25} & K_{26} \\ K_{31} & K_{32} & K_{33} & K_{34} & K_{35} & K_{36} \end{pmatrix}$$

Class C_S (10)

$$\begin{pmatrix} K_{11} & K_{12} & K_{13} & 0 & 0 & K_{16} \\ K_{21} & K_{22} & K_{23} & 0 & 0 & K_{26} \\ 0 & 0 & 0 & K_{34} & K_{35} & 0 \end{pmatrix}$$

Class C_2 (8)

$$\begin{pmatrix} 0 & 0 & 0 & K_{14} & K_{15} & 0 \\ 0 & 0 & 0 & K_{24} & K_{25} & 0 \\ K_{31} & K_{32} & K_{33} & 0 & 0 & K_{36} \end{pmatrix}$$

Class C_{2v} (5)

$$\begin{pmatrix} 0 & 0 & 0 & 0 & K_{15} & 0 \\ 0 & 0 & 0 & K_{24} & 0 & 0 \\ K_{31} & K_{32} & K_{33} & 0 & 0 & 0 \end{pmatrix}$$

Class D_2 (3)

$$\begin{pmatrix} 0 & 0 & 0 & K_{14} & 0 & 0 \\ 0 & 0 & 0 & 0 & K_{25} & 0 \\ 0 & 0 & 0 & 0 & 0 & K_{36} \end{pmatrix}$$

Classes C_4 *and* C_6 (4)

$$\begin{pmatrix} 0 & 0 & 0 & K_{14} & K_{15} & 0 \\ 0 & 0 & 0 & K_{15} & -K_{14} & 0 \\ K_{31} & K_{31} & K_{33} & 0 & 0 & 0 \end{pmatrix}$$

Class S_4 (4)

$$\begin{pmatrix} 0 & 0 & 0 & K_{14} & K_{15} & 0 \\ 0 & 0 & 0 & -K_{15} & K_{14} & 0 \\ K_{31} & -K_{31} & 0 & 0 & 0 & K_{36} \end{pmatrix}$$

Classes C_{4v}, C_{6v} *and* $C_{\infty v}$ (3)

$$\begin{pmatrix} 0 & 0 & 0 & 0 & K_{15} & 0 \\ 0 & 0 & 0 & K_{15} & 0 & 0 \\ K_{31} & K_{31} & K_{33} & 0 & 0 & 0 \end{pmatrix}$$

Class D_{2d} (2)

$$\begin{pmatrix} 0 & 0 & 0 & K_{14} & 0 & 0 \\ 0 & 0 & 0 & 0 & K_{14} & 0 \\ 0 & 0 & 0 & 0 & 0 & K_{36} \end{pmatrix}$$

Classes D_4 *and* D_6 (1)

$$\begin{pmatrix} 0 & 0 & 0 & K_{14} & 0 & 0 \\ 0 & 0 & 0 & 0 & -K_{14} & 0 \\ 0 & 0 & 0 & 0 & 0 & 0 \end{pmatrix}$$

Class C_3 (6)

$$\begin{pmatrix} K_{11} & -K_{11} & 0 & K_{14} & K_{15} & -\sqrt{2}K_{22} \\ -K_{22} & K_{22} & 0 & K_{15} & -K_{14} & -\sqrt{2}K_{11} \\ K_{31} & K_{31} & K_{33} & 0 & 0 & 0 \end{pmatrix}$$

Class C_{3v} (4)

$$\begin{pmatrix} 0 & 0 & 0 & 0 & K_{15} & -\sqrt{2}K_{22} \\ -K_{22} & K_{22} & 0 & K_{15} & 0 & 0 \\ K_{31} & K_{31} & K_{33} & 0 & 0 & 0 \end{pmatrix}$$

Class D_3 (2)

$$\begin{pmatrix} K_{11} & -K_{11} & 0 & K_{14} & 0 & 0 \\ 0 & 0 & 0 & 0 & -K_{14} & -\sqrt{2}K_{11} \\ 0 & 0 & 0 & 0 & 0 & 0 \end{pmatrix}$$

Class C_{3h} (2)

$$\begin{pmatrix} K_{11} & -K_{11} & 0 & 0 & 0 & -\sqrt{2}K_{22} \\ -K_{22} & K_{22} & 0 & 0 & 0 & -\sqrt{2}K_{11} \\ 0 & 0 & 0 & 0 & 0 & 0 \end{pmatrix}$$

Class D_{3h} (1)

$$\begin{pmatrix} 0 & 0 & 0 & 0 & 0 & -\sqrt{2}K_{22} \\ -K_{22} & K_{22} & 0 & 0 & 0 & 0 \\ 0 & 0 & 0 & 0 & 0 & 0 \end{pmatrix}$$

Classes T *and* T_d (1)

$$\begin{pmatrix} 0 & 0 & 0 & K_{14} & 0 & 0 \\ 0 & 0 & 0 & 0 & K_{14} & 0 \\ 0 & 0 & 0 & 0 & 0 & K_{14} \end{pmatrix}$$

and then for **K**:

$$\mathbf{K} = \begin{pmatrix} 0 & 0 & 0 & 0 & K_{15} & 0 \\ 0 & 0 & 0 & K_{15} & 0 & 0 \\ K_{31} & K_{31} & K_{33} & 0 & 0 & 0 \end{pmatrix} \qquad (7\text{–}11)$$

where

$$K_{15} = K_{\mathrm{E}},\ K_{33} = K_{\mathrm{A},2} \quad \text{and} \quad K_{31} = K_{\mathrm{A},1}/\sqrt{2} \qquad (7\text{–}12)$$

The same result is obtained for the group C_{6v}.

The striking contrast between Eq. (11) for C_{4v} or C_{6v} and Eq. (6) for D_4 or D_6 is due, on the one hand, to the difference in orientational matching of symmetry coordinates for the E irrep in the two cases and, on the other hand, to the coupling of **X** and **Y** components belonging to the A_1 irrep for the C_{nv} groups.

The group C_{3v} is similar to the above example except that α_6 and $\alpha_1 - \alpha_2$ also belong to the E irrep with orientations similar to x and y. Thus we have repeats in both the A and E irreps on the part of the $\bar{\mathbf{X}}$-coordinates. Keeping the same symmetry coordinates as in Eq. (7), we obtain

$$\bar{\mathbf{K}} = \begin{pmatrix} 0 & 0 & 0 & 0 & K_{\mathrm{E},2} & K_{\mathrm{E},1} \\ 0 & K_{\mathrm{E},1} & 0 & K_{\mathrm{E},2} & 0 & 0 \\ K_{\mathrm{A},1} & 0 & K_{\mathrm{A},2} & 0 & 0 & 0 \end{pmatrix} \qquad (7\text{–}13)$$

so that

$$\mathbf{K} = \begin{pmatrix} 0 & 0 & 0 & 0 & K_{15} & -\sqrt{2}K_{22} \\ -K_{22} & K_{22} & 0 & K_{15} & 0 & 0 \\ K_{31} & K_{31} & K_{33} & 0 & 0 & 0 \end{pmatrix} \qquad (7\text{–}14)$$

where

$$K_{15} = K_{\mathrm{E},2}, \quad -K_{22} = K_{\mathrm{E},1}/\sqrt{2} \qquad (7\text{–}15)$$

while K_{31} and K_{33} remain as in Eq. (12).

Although the results for lower symmetry groups become more complex (i.e. the number of independent coefficients increases), the method of treatment remains the same. We will therefore refer the reader to the complete Table 7–1 and to the problems for further results.

7–1–2 Application to piezoelectricity: quartz and PZT

The piezoelectric effect is the generation of an electric displacement (or polarization) by a stress (direct effect), or the production of strain by an electric field (converse effect). (See Eqs. (1–10) and (1–16).) The piezoelectric coefficients may be given in three-index notation, d_{pqr}, or in two-index engineering notation d^{e}_{ij}. (The latter are the same as the two-index K_{ij} coefficient in Table 7–1, except that all factors of $\sqrt{2}$ must be converted to factors of 2.)

Of all crystals used for their piezoelectric property, α-quartz dominates by far. It is used in oscillators to provide stable frequencies, as well as for generating and receiving mechanical vibrations and for high-selectivity filters. Until World War II, only natural quartz crystals were in use. Due to shortages of natural crystals, methods of synthetic growth were developed at that time, leading to very-high-quality crystals, superior to the natural crystals, through control of impurities. Today, quartz is probably the most widely used synthetic crystalline material, second only to silicon.

The crystal structure of α-quartz is that of point group D_3. It therefore (see Table 7–1) has two independent piezoelectric coefficients d_{11} and d_{14} with $d_{11} = -d_{12}$. The values of these coefficients are $d^{e}_{11} = 2.25$ and $d^{e}_{14} = -0.85$ in 10^{-12} coul./newton.

Figure 7–1 shows some of the orientations of quartz crystals used as transducers. The simplest excitation is that of an X-cut crystal, Fig. 7–1(a), in which the electrodes are applied on a surface perpendicular to

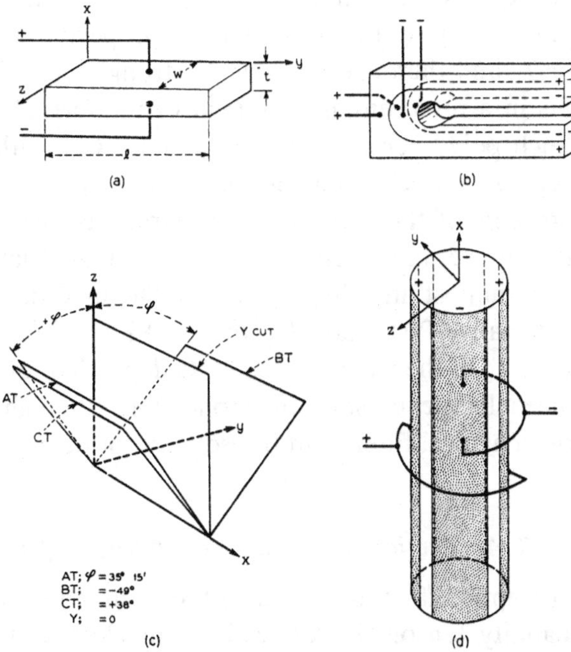

Fig. 7–1. Some orientations of quartz crystals for use as transducers: (*a*) longitudinal, (*b*) tuning fork, (*c*) shear vibrating, (*d*) torsional. (After Mason, 1958.)

the x-axis. In this manner, one can obtain thickness vibrations of plates, utilizing the d_{11} coefficient, or lengthwise vibrations along the y-axis due to the coefficient d_{12} ($= -d_{11}$). Piezoelectric excitation of shear vibration can also be produced as a consequence of the d_{14} coefficient. It is also possible to set up a 'tuning-fork crystal' as shown in Fig. 7–1(b). Slabs cut as in Fig. 7–1(c) may be used for shear vibrations, the simplest being the Y-cut. Since most applications of quartz resonators require a low-temperature coefficient of the frequency, as well as avoidance of the interfering effects of other modes of motion, considerable work has been done on crystal plates with various orientations to the standard crystal axes. So-called AT, BT, CT, DT and GT cuts (some of which are shown in Fig. 7–1) have very small frequency changes over a relatively wide range of temperature near room temperature. Of these, the AT and GT are most widely used to control the frequency of oscillators.

It is also possible to excite flexural and torsional vibrations by proper orientation of crystals and electrodes (e.g. see Fig. 7–1(d)).

Recently, certain ferroelectric ceramics have become strong competitors of quartz for some piezoelectric applications. These ceramics are not used as single crystals but, rather, as polycrystals. The way that they achieve the piezoelectric property is by a process called 'poling'. Since the materials are ferroelectrics (see Section 6–2), they possess a Curie temperature, T_c, above which they lose the ferroelectric property. The ceramic is heated to above T_c and then cooled under a strong electric field, **E**. This procedure results in a preferential alignment of the ferroelectric domains parallel to **E**. Such poled ceramics have symmetry $C_{\infty v}$ with three piezoelectric constants (same as C_{4v}), namely, $d_{31} = d_{32}$, d_{33} and $d_{15} = d_{24}$. The piezoelectric constants of these ceramics are large (as much as 100 times those of quartz). The ceramics also have the advantage that they may be readily prepared in almost any shape or form. The materials used include $BaTiO_3$ and, especially, $PbTi_xZr_{1-x}O_3$ (called PZT). PZT, which has a Curie temperature T_c in the range 328–370 °C depending on composition, has replaced piezoelectric crystals in applications to high-power sonar systems, electromechanical filters and transducers for delay lines.

7–1–3 Application to the linear electro-optic effect

The linear electro-optic effect is related to the change in the components of the impermeability tensor $\boldsymbol{\beta}$ produced by an electric field. (See Eq. (1–31).) The coefficients may be expressed in three-index notation r_{ijk} (i, j, $k = 1, 2, 3$) or in two-index form r_{mk} ($m = 1, \ldots, 6$) in which the first two indices are converted into a six-vector in the manner of Eq. (4–4). Finally, they may be written in two-index engineering form r^{e}_{mk} in which the conversion is made in the same manner of the stresses in Section 1–4. The two-index electro-optic coefficients r_{mk} for all crystal symmetries may be obtained from Table 7–1 simply by transposing the 3×6 matrices into 6×3 matrices. On the other hand, to obtain the engineering coefficients requires dropping all factors of $\sqrt{2}$.

The description of the electro-optic effect is given in terms of the indicatrix (or index ellipsoid) as described by Eq. (1–28). The reader is reminded that the lengths of the principal axes of this ellipsoid are n_i, where the n_i are the principal refractive indices. (See Eq. (1–30).) With the introduction of the electro-optic coefficients the indicatrix becomes

$$\left[\frac{1}{n_i^2} + r_{1k}E_k\right]x^2 + \left[\frac{1}{n_2^2} + r_{2k}E_k\right]y^2 + \left[\frac{1}{n_3^2} + r_{3k}E_k\right]z^2 + 2yzr_{4k}E_k + 2xzr_{5k}E_k + 2xyr_{6k}E_k = 1 \tag{7–16}$$

where all the r_{ik} coefficients are the two-index engineering quantities (the super e being dropped for simplicity). If light travels parallel or perpendicular to **E**, we are dealing, respectively, with the longitudinal or transverse electro-optic effect.

First, let us consider utilizing a crystal of T_d symmetry, for example GaAs. In this case, there is only one coefficient, namely r_{41}, and the longitudinal effect is appropriate. In the absence of an electric field this (cubic) crystal is optically isotropic with a single refractive index n. With an applied field E_3 along the z-axis, the indicatrix becomes, simply

$$(x^2 + y^2 + z^2)/n^2 + 2xyr_{41}E_3 = 1 \tag{7–17}$$

since $r_{63} = r_{41}$ (see Table 7–1). This ellipsoid can be rotated into principal axes by taking new axes x' and y' at ±45° to the x- and y-axes while keeping z unchanged. The indices of refraction along the new principal axes are

$$\left.\begin{aligned} n_{x'} &= \mathrm{n} - (1/2)n^3 r_{41}E_3 \\ n_{y'} &= n + (1/2)n^3 r_{41}E_3 \\ n_z &= n \end{aligned}\right\} \tag{7–18}$$

where we have assumed that $n^2 r_{41}E_3 \ll 1$. Thus, application of electric field E_3 has caused the cubic (T_d) crystal to become birefringent. Because of this birefringence, the crystal may function as a wave plate in a modulator, as shown in Fig. 7–2. Here the input optical wave, traveling along the z-direction, is polarized along x. As a result of Eq. (18), there

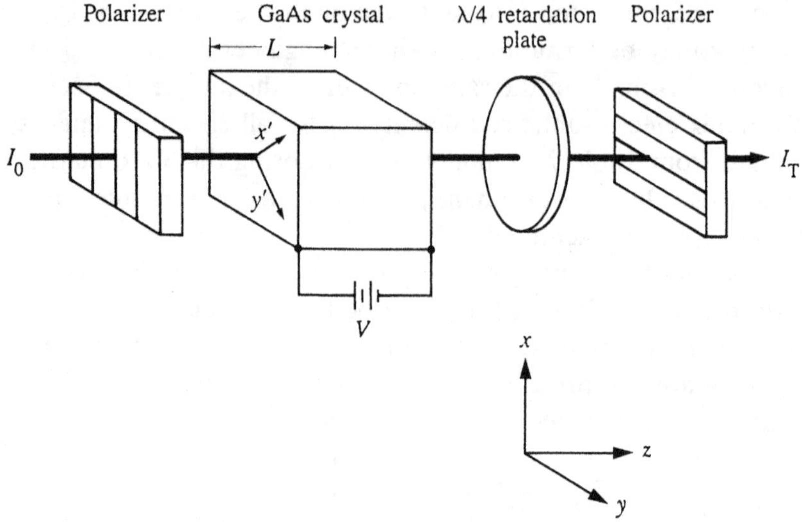

Fig. 7–2. Intensity modulator based on the longitudinal electro-optic effect, employing a crystal such as GaAs. (After Wiesenfeld, 1990.)

will be a phase difference between the y'- and x'-components of the wave, called the retardation Γ, which is given by

$$\Gamma = 2\pi(n_{y'} - n_{x'})L/\lambda = 2\pi n^3 r_{41} V/\lambda \tag{7–19}$$

where λ is the vacuum wavelength, V the applied voltage and L the length of the crystal in the z-direction. In the absence of the quarter-wave ($\lambda/4$) plate in the diagram, the optical intensity I_T through the output polarizer will be

$$I_T = I_0 \sin^2(\Gamma/2) \tag{7–20}$$

where I_0 is the incident intensity. This transmitted intensity reaches a maximum when $\Gamma = \pi$, under which circumstances, the addition of the two waves produces a plane wave whose direction of polarization is parallel to that of the analyzer. The appropriate voltage required to achieve this maximum is called V_π, and its measurement clearly allows the determination of r_{41}. In order to obtain a linear response to an applied voltage, however, a static $\lambda/4$ retardation plate is inserted, as shown in Fig. 7–2. This wave plate serves to optically bias the modulator to $\Gamma = \pi/2$, where it is most sensitive to changes in applied voltage.

It is worth noting, from Eq. (19), that the quantity $n^3 r_{\text{eff}}$ (r_{eff} being the effective electro-optic coefficient) is a measure of the electro-optic figure of merit.

Until now, we have dealt with a longitudinal electro-optic effect. With uniaxial crystals, it is also possible to use the transverse effect. In particular, crystals of $LiNbO_3$ and $LiTaO_3$ (point group C_{3v}) have been widely used. These crystals begin with a natural birefringence, with optic axis parallel to z and with n_o and n_e the ordinary and extraordinary refractive indices (see Fig. 6–2). In this case, the principal axes are unaltered by a field E_3; however, the indices of refraction are altered, and given by:

$$\left.\begin{aligned} n_x = n_y &= n_o - 1/2 n_o^3 r_{13} E_3 \\ n_z &= n_e - 1/2 n_e^3 r_{33} E_3 \end{aligned}\right\} \qquad (7\text{–}21)$$

involving the non-zero electro-optic coefficients r_{13} and r_{33} (see Table 7–1 for C_{3v} symmetry). If the light beam is propagated along the y-axis and is given an input polarization at 45° to the x- and z-axes, the phase difference or retardation (after removing the effect of the initial birefringence with a compensator) is given by

$$\Gamma = \pi(n_e^3 r_{33} - n_o^3 r_{13}) L_2 V/\lambda d \qquad (7\text{–}22)$$

where d is the crystal thickness and L_2 is its length in the y-direction.

The value of the pertinent electro-optic coefficients for $LiNbO_3$ are $r_{13} = 10$ pm/V and $r_{33} = 32.6$ pm/V, while the refractive indices are given in Table 6–2. These values give rise to a particularly high figure of merit, which makes this material very attractive for modern electro-optic applications. Other materials studied earlier include KDP (KH_2PO_4) and ADP ($NH_4H_2PO_4$), both of D_{2d} symmetry, as well as ZnS and CuCl, both of T_d symmetry.

There are numerous applications for electro-optic materials. The modulator shown in Fig. 7–2 can be used for electro-optic sampling, which is an optical probing technique for the measurement of signals from high-speed electronic devices, for example those that are part of an integrated circuit (Wiesenfeld, 1990). Another application is for photorefractive materials, involving a change in local refractive index under inhomogeneous illumination. This application involves the combination of two phenomena: first, free carriers are generated by illumination, which is a photoconductivity effect. The inhomogeneity of these carriers then generates a space-charge field that causes a correlated index change as a result of the electro-optic effect. Photorefractive materials can be used for optical image processing. (See Günter, 1987.)

Other applications of the linear electro-optic effect are discussed by Narasimhamurty (1981). They include light modulators, light shutters and function converters.

7–2 Non-symmetric tensors of rank 3

It seems appropriate to consider next the case of matter tensors of type $T(3)$, which can be obtained by the coupling of a vector force to a $T(2)$ response, or vice versa. Since the $T(2)$ quantity possesses nine components, the maximum number of K components in this case is 27, that is, $\mathbf{K}$ is a 3×9 (or 9×3) matrix.

The reader will note that no examples of this kind are listed in Tables 1–1 and 2–1, implying that there is no widely studied phenomenon that falls in the $T(3)$ category. This does not mean that such phenomena do not exist. In fact, there are higher-order effects of this type, an example of which would be the effect of an electric field on any $T(2)$ tensor property, such as the Hall effect.

To deal with such a problem group theoretically, we must convert the two-index $T(2)$ quantity into a single-index nine-vector. This we accomplish by first separating the components into symmetric and antisymmetric linear combinations. Thus, consider that $\mathbf{X}$ in Eq. (1) is of tensor character $T(2)$ having components X_{ij}. In the usual way, we separate $\mathbf{X}$ into symmetric and antisymmetric second-rank tensors, $\mathbf{X}^{\mathrm{S}}$ and $\mathbf{X}^{\mathrm{aS}}$, respectively, such that

$$\mathbf{X}^{\mathrm{S}} + \mathbf{X}^{\mathrm{aS}} = \mathbf{X} \tag{7–23}$$

This is done by writing

$$\left.\begin{aligned} X_{ij}^{\mathrm{S}} &= X_{ij} && (i = j) \\ X_{ij}^{\mathrm{S}} &= 1/2(X_{ij} + X_{ji}) && (i \neq j) \end{aligned}\right\} \tag{7–24}$$

and

$$\left.\begin{aligned} X_{ij}^{\mathrm{aS}} &= 0 && (i = j) \\ X_{ij}^{\mathrm{aS}} &= 1/2(X_{ij} - X_{ji}) && (i \neq j) \end{aligned}\right\} \tag{7–25}$$

The advantage of this separation is that the six X^{S} components and the three X^{aS} components will never mix under a symmetry operation, the first set transforming as Γ_{α} and the second as $\Gamma_{\mathbf{R}}$. (See Section 3–4.) In other words, this separation into symmetric and antisymmetric components is a first step toward obtaining the symmetry coordinates of the nine-vector. However, we already know how to form a six-vector out of the X^{S} components by the Weyl normalization (Eq. (4–4)), and an axial three-vector out of the X^{aS} components. Thus, we form the single-index nine-vector as follows:

$$\left.\begin{aligned} &\text{Nine-vector:} && X_1, X_2, X_3, X_4, X_5, X_6, X_7, X_8, X_9 \\ &\text{Tensor:} && X_{11}, X_{22}, X_{33}, (X_{23} + X_{32})/\sqrt{2}, (X_{31} + X_{13})/\sqrt{2}, \\ &&& (X_{12} + X_{21})/\sqrt{2}, (X_{23} - X_{32})/\sqrt{2}, \\ &&& (X_{31} - X_{13})/\sqrt{2}, (X_{12} - X_{21})/\sqrt{2} \end{aligned}\right\} \tag{7–26}$$

This nine-vector is then the basis of a rep that transforms as $\Gamma_\alpha + \Gamma_{\mathbf{R}}$. Inverting this transformation, in order to pass from one-index to two-index quantities, we have

$$X_{23} = (X_4 + X_7)/\sqrt{2}; \quad X_{32} = (X_4 - X_7)/\sqrt{2} \tag{7–27}$$

and similarly for components 31 and 13 from $X_5 \pm X_8$ and components 12 and 21 from $X_6 \pm X_9$.

With this background, we are prepared to handle the case of matter tensors of type $T(3)$. However, for reasons that will be explained later (Section 7–4), we shall first move on to axial tensors of rank 3. In any case, Eq. (27) provides the basis for the conversion of **K** from two-index to three-index notation, as follows:

$$\left.\begin{aligned} K_{ij} &= K_{ijj} \quad \text{for } j = 1, 2, 3. \\ K_{i23} &= (K_{i4} + K_{i7})/\sqrt{2}; \quad K_{i32} = (K_{i4} - K_{i7})/\sqrt{2} \end{aligned}\right\} \tag{7–28}$$

and similarly for components $i31$ and $i13$ obtained from $K_{i5} \pm K_{i8}$ and $i12$ and $i21$ obtained from $K_{i6} \pm K_{i9}$.

The only difference in the case of 9×3 (transposed) two-index matrices is that the index i now appears in the last position, while the remaining indices are treated as in Eqs. (28).

7–3 Axial tensors of rank 3

It would seem logical to begin with partly symmetric axial tensors obtained from the coupling between an axial vector, $T(1)^{\text{ax}}$, and a second-rank symmetric tensor, $T_{\text{S}}(2)$. An example is the case of piezomagnetism in Table 1–1 (although this is a 'special magnetic property'). It turns out, however, that we need not work out this tensor type separately, since it falls out as a special case of the more general matter tensor $T(3)^{\text{ax}}$ obtained by coupling of a $T(1)^{\text{ax}}$ and a non-symmetric $T(2)$. An example is the Nernst tensor of Eq. (2–28) and Table 2–1. Again, to be specific, let us take **Y** in the form of the axial vector and **X** as the $T(2)$ tensor, expressed as a nine-vector as in Eq. (26), so that **K** is a 3×9 matrix. (In fact, the Nernst tensor reverses the roles of **X** and **Y** and so requires a transposed **K** in a 9×3 form.) The nine-vector **X** then transforms as $\Gamma_\alpha + \Gamma_{\mathbf{R}}$ while the axial vector **Y** goes as $\Gamma_{\mathbf{R}}$. Clearly, there will always be common irreps of **X** and **Y**, that is, **K** is never identically zero.

It is noteworthy that the structure of this tensor subdivides itself according to the Laue groups (see page 97).

For the higher symmetry cubic groups (O_h, T_d and O), there is no common irrep of Γ_α and $\Gamma_{\mathbf{R}}$, but only of $\Gamma_{\mathbf{R}}$ with itself (through a T-type irrep). We therefore obtain **K** as in Tables 7–2, with a single constant K_{17} given by

$$K_{17} = K_{\mathrm{T}} \qquad (7\text{–}29)$$

The conversion to three-index notation follows from Eq. (28) and the result is given in Table 7–3.

In the lower cubic groups ($\mathrm{T_h}$ and T), α_4, α_5 and α_6 belong to the same T-irrep as R_x, R_y and R_z. Stated differently, the irrep T is repeated in the decomposition of the rep of **X**. Accordingly, **K** now has two constants, K_{14} and K_{17} (see Table 7–2), where

$$K_{14} = K_{\mathrm{T},1} \quad \text{and} \quad K_{17} = K_{\mathrm{T},2} \qquad (7\text{–}30)$$

This result can be converted to three-index notation using Eq. (28) to obtain, for the non-zero coefficients,

$$K_{123} = K_{231} = K_{312} = (K_{14} + K_{17})/\sqrt{2} \qquad (7\text{–}31)$$

and

$$K_{132} = K_{213} = K_{321} = (K_{14} - K_{17})/\sqrt{2} \qquad (7\text{–}32)$$

The corresponding three-index **K** matrix is given in Table 7–3.

Turning now to the upper hexagonal and tetragonal groups, we find that R_z stands alone in an A irrep while $(R_x, -R_y)$ is matched in orientation by (α_4, α_5) in a repeated E irrep. Thus **K** is obtained as in Table 7–2 with

$$K_{39} = K_{\mathrm{A}}; \quad K_{14} = K_{\mathrm{E},1}; \quad K_{17} = K_{\mathrm{E},2} \qquad (7\text{–}33)$$

Converting to three-index notation, we obtain

$$K_{123} = -K_{213} = (K_{14} + K_{17})/\sqrt{2} \qquad (7\text{–}34)$$

$$K_{132} = -K_{231} = (K_{14} - K_{17})/\sqrt{2} \qquad (7\text{–}35)$$

$$K_{312} = -K_{321} = K_{39}/\sqrt{2} \qquad (7\text{–}36)$$

The upper trigonal systems are different. Again, R_z appears alone as an A irrep but now $(R_x, -R_y)$, (α_4, α_5) and $(\alpha_1 - \alpha_2, \alpha_6)$ all belong to the E irrep. Thus among the symmetry coordinates of **X**, irrep E is triply repeated. If we then choose as symmetry coordinates $\bar{X}_1, \ldots, \bar{X}_9$ the set: $(X_1 + X_2)/\sqrt{2}$, $(X_1 - X_2)/\sqrt{2}$, $X_3, \ldots, X_9$, we obtain for the matrix $\bar{\mathbf{K}}$

$$\bar{\mathbf{K}} = \begin{pmatrix} 0 & K_{\mathrm{E},3} & 0 & K_{\mathrm{E},2} & 0 & 0 & K_{\mathrm{E},1} & 0 & 0 \\ 0 & 0 & 0 & 0 & -K_{\mathrm{E},2} & -K_{\mathrm{E},3} & 0 & K_{\mathrm{E},1} & 0 \\ 0 & 0 & 0 & 0 & 0 & 0 & 0 & 0 & \mathrm{K_A} \end{pmatrix} \qquad (7\text{–}37)$$

from which the **K** matrix in Table 7–2 readily follows, with

$$\left.\begin{array}{l} K_{11} = -K_{26}/\sqrt{2} = K_{\mathrm{E},3}/\sqrt{2} \\ K_{14} = K_{\mathrm{E},2}; \quad K_{17} = K_{\mathrm{E},1}; \quad K_{39} = K_{\mathrm{A}} \end{array}\right\} \qquad (7\text{–}38)$$

For the three-index coefficients, we obtain

$$K_{111} = -K_{122} = -K_{212} = -K_{221} = K_{11} \qquad (7\text{–}39)$$

while Eqs. (34), (35) and (36) remain valid for this case, too.

In this present study, we can see the benefit of choosing the pair (α_4, α_5) to fix the E irrep orientation in the S-C-T tables, rather than the pair (x, y). By making this choice, it becomes readily apparent that the matching of α-coefficients and R-coefficients belonging to the E irrep is the same for all upper hexagonal and tetragonal groups (e.g. D_n as well as C_{nv}), and similarly for the upper trigonal groups. If (x, y) had been used to fix the orientation, the same results would, of course, have been obtained, but it would have been somewhat more difficult to see the commonality of upper uniaxial groups from the S-C-T tables.

To illustrate the procedure for obtaining **K** in a case in which complex irreps appear, let us now turn to the lower hexagonal and tetragonal groups. In these cases, belonging to irrep A (or A_g) we find R_z as well as $\alpha_1 + \alpha_2$ and α_3, while under $\tilde{E}$ we have $R_x \mp iR_y$ and $\alpha_4 \pm i\alpha_5$. In view of what was worked out in Chapter 4 for the $\tilde{E}$ irreps (Eqs. (4–24) to (4–26)), we need not use complex quantities for the symmetry coordinates associated with the $\tilde{E}$ irrep. Accordingly, we may make the same choice of symmetry coordinates as for the upper groups: namely, for $\bar{Y}_1$, $\bar{Y}_2$ and $\bar{Y}_3$ the quantities Y_1, Y_2 and Y_3, and for the nine symmetry coordinates $\bar{X}_1$ to $\bar{X}_9$: $(X_1 + X_2)/\sqrt{2}$, $(X_1 - X_2)/\sqrt{2}$, X_3, X_4, X_5, X_6, X_7, X_8 and X_9, in that order. In these terms, the matrix $\bar{\mathbf{K}}$ can be written directly as

$$\bar{\mathbf{K}} = \begin{pmatrix} 0 & 0 & 0 & \tilde{E},1^{re} & \tilde{E},1^{im} & 0 & \tilde{E},2^{re} & \tilde{E},2^{im} & 0 \\ 0 & 0 & 0 & \tilde{E},1^{im} & -\tilde{E},1^{re} & 0 & -\tilde{E},2^{im} & \tilde{E},2^{re} & 0 \\ A,2 & 0 & A,3 & 0 & 0 & 0 & 0 & 0 & A,1 \end{pmatrix} \qquad (7\text{–}40)$$

Here we have simplified the notation by showing only the sign, the subscript (on the line) and the superscript for each entry, except for those that are zero. Note that $\bar{X}_2$ and $\bar{X}_6$ are not coupled to any $\bar{Y}$-coordinates.

The next step is to convert to the **K** matrix. This can be done, in a formal way, using Eqs. (4–9) and (4–10), so that $\mathbf{K} = \mathbf{T}\bar{\mathbf{K}}\mathbf{S}^{\dagger}$, or, more simply, by direct comparison of coefficients. Either way, we obtain the matrix **K** given in Table 7–2, with

$$\left.\begin{aligned} &K_{14} = K^{re}_{\tilde{E},1}; \quad K_{15} = K^{im}_{\tilde{E},1}; \quad K_{17} = K^{re}_{\tilde{E},2}; \quad K_{18} = K^{im}_{\tilde{E},2} \\ &K_{31} = K_{A,2}/\sqrt{2}; \quad K_{33} = K_{A,3}; \quad K_{39} = K_{A,1} \end{aligned}\right\} \qquad (7\text{–}41)$$

Finally, the three-index coefficients may be obtained with the aid of

Eqs. (28) as

$$K_{311} = K_{322} = K_{31}; \quad K_{333} = K_{33}; \quad K_{312} = -K_{321} = K_{39}/\sqrt{2} \quad (7\text{–}42)$$

$$K_{123} = -K_{213} = (K_{14} + K_{17})/\sqrt{2} \quad (7\text{–}43)$$

$$K_{132} = -K_{231} = (K_{14} - K_{17})/\sqrt{2}; \quad K_{113} = K_{223} = (K_{15} - K_{18})/\sqrt{2} \quad (7\text{–}44)$$

$$K_{131} = K_{232} = (K_{15} + K_{18})/\sqrt{2} \quad (7\text{–}45)$$

In a similar way, the two-index and three-index **K** matrices may be obtained for other crystal groups. The results are presented in Tables 7–2 and 7–3.

Let us now return to the case (discussed at the beginning of this section) of an axial **K** tensor obtained when an axial vector couples to a $T_S(2)$ tensor. In this case **X** (taken as the $T_S(2)$ tensor) has only six components instead of nine, and **K** is a 3×6 matrix. It is clear that the two-index **K** matrices in this case are just the first six columns of the matrices that appear in Table 7–2. Thus, there is no need to work out separately the partly symmetric $T(3)^{ax}$ tensor. On the other hand, for practical notation, these tensors need to be put into engineering notation, as, for example, when **X** is a stress. In this case, the conversion is the same as that given in Eq. (3) for the piezoelectric coefficients.

It is also worth noting that, in view of the theorem of non-mixing of symmetric and antisymmetric coordinates under symmetry operations (Problem 3–13), the last three columns of Table 7–2 are just the relations between $T(1)^{ax}$ and $T(1)^{ax}$, forming a 3×3 $T(2)$ tensor that we have already dealt with in Chapter 6.

7–4 Polar tensors of rank 3 revisited

This type of non-symmetric tensor T(3) has already been discussed earlier in this chapter, Section 7–2, but we did not work out the tensor in detail for different symmetries. Now, it is apparent that we can create such 3×9 **K** matrices, for the coupling of a $T(2)$ force to a $T(1)$ response, simply by putting together the first six columns obtained from the case of a 3×6 partly-symmetric $T(3)$ tensor (Table 7–1), with three additional columns of the $T(2)^{ax}$ tensor, Table 6–3. (However, the latter represents the coupling of $T(1)^{ax}$ to $T(1)$ so that the required 3×3 matrix must be the *transpose* of the one given in Table 6–3.) Accordingly, there is no need to rework tables for the present case. Conversion to a three-index notation would take place as shown in Eq. (28).

Table 7–2. *Forms of a $T(3)^{ax}$ tensor in two-index notation for various crystal symmetries*

Triclinic (27 constants)

$$\begin{pmatrix} K_{11} & K_{12} & K_{13} & K_{14} & K_{15} & K_{16} & K_{17} & K_{18} & K_{19} \\ K_{21} & K_{22} & K_{23} & K_{24} & K_{25} & K_{26} & K_{27} & K_{28} & K_{29} \\ K_{31} & K_{32} & K_{33} & K_{34} & K_{35} & K_{36} & K_{37} & K_{38} & K_{39} \end{pmatrix}$$

Monoclinic (13)

$$\begin{pmatrix} 0 & 0 & 0 & K_{14} & K_{15} & 0 & K_{17} & K_{18} & 0 \\ 0 & 0 & 0 & K_{24} & K_{25} & 0 & K_{27} & K_{28} & 0 \\ K_{31} & K_{32} & K_{33} & 0 & 0 & K_{36} & 0 & 0 & K_{39} \end{pmatrix}$$

Orthorhombic (6)

$$\begin{pmatrix} 0 & 0 & 0 & K_{14} & 0 & 0 & K_{17} & 0 & 0 \\ 0 & 0 & 0 & 0 & K_{25} & 0 & 0 & K_{28} & 0 \\ 0 & 0 & 0 & 0 & 0 & K_{36} & 0 & 0 & K_{39} \end{pmatrix}$$

Trigonal: upper (4)

$$\begin{pmatrix} K_{11} & -K_{11} & 0 & K_{14} & 0 & 0 & K_{17} & 0 & 0 \\ 0 & 0 & 0 & 0 & -K_{14} & -\sqrt{2}K_{11} & 0 & K_{17} & 0 \\ 0 & 0 & 0 & 0 & 0 & 0 & 0 & 0 & K_{39} \end{pmatrix}$$

Trigonal: lower (9)

$$\begin{pmatrix} K_{11} & -K_{11} & 0 & K_{14} & -K_{15} & -\sqrt{2}K_{22} & K_{17} & K_{18} & 0 \\ -K_{22} & K_{22} & 0 & K_{15} & -K_{14} & -\sqrt{2}K_{11} & -K_{18} & K_{17} & 0 \\ K_{31} & K_{31} & K_{33} & 0 & 0 & 0 & 0 & 0 & K_{39} \end{pmatrix}$$

Tetragonal and hexagonal; upper (3)

$$\begin{pmatrix} 0 & 0 & 0 & K_{14} & 0 & 0 & K_{17} & 0 & 0 \\ 0 & 0 & 0 & 0 & -K_{14} & 0 & 0 & K_{17} & 0 \\ 0 & 0 & 0 & 0 & 0 & 0 & 0 & 0 & K_{39} \end{pmatrix}$$

Tetragonal and hexagonal: lower (7)

$$\begin{pmatrix} 0 & 0 & 0 & K_{14} & K_{15} & 0 & K_{17} & K_{18} & 0 \\ 0 & 0 & 0 & K_{15} & -K_{14} & 0 & -K_{18} & K_{17} & 0 \\ K_{31} & K_{31} & K_{33} & 0 & 0 & 0 & 0 & 0 & K_{39} \end{pmatrix}$$

Cubic: upper (1)

$$\begin{pmatrix} 0 & 0 & 0 & 0 & 0 & 0 & K_{17} & 0 & 0 \\ 0 & 0 & 0 & 0 & 0 & 0 & 0 & K_{17} & 0 \\ 0 & 0 & 0 & 0 & 0 & 0 & 0 & 0 & K_{17} \end{pmatrix}$$

Cubic: lower (2)

$$\begin{pmatrix} 0 & 0 & 0 & K_{14} & 0 & 0 & K_{17} & 0 & 0 \\ 0 & 0 & 0 & 0 & K_{14} & 0 & 0 & K_{17} & 0 \\ 0 & 0 & 0 & 0 & 0 & K_{14} & 0 & 0 & K_{17} \end{pmatrix}$$

Table 7–3. *Forms of a* T*(3)*^{ax} *tensor in three-index notation, giving only indices and sign (except where values are zero)*

Triclinic (27 constants)

$$\begin{pmatrix} 111 & 121 & 131 & 112 & 122 & 132 & 113 & 123 & 133 \\ 211 & 221 & 231 & 212 & 222 & 232 & 213 & 223 & 233 \\ 311 & 321 & 331 & 312 & 322 & 332 & 313 & 323 & 333 \end{pmatrix}$$

Monoclinic (13)

$$\begin{pmatrix} 0 & 0 & 131 & 0 & 0 & 132 & 113 & 123 & 0 \\ 0 & 0 & 231 & 0 & 0 & 232 & 213 & 223 & 0 \\ 311 & 321 & 0 & 312 & 322 & 0 & 0 & 0 & 333 \end{pmatrix}$$

Orthorhombic (6)

$$\begin{pmatrix} 0 & 0 & 0 & 0 & 0 & 132 & 0 & 123 & 0 \\ 0 & 0 & 231 & 0 & 0 & 0 & 213 & 0 & 0 \\ 0 & 321 & 0 & 312 & 0 & 0 & 0 & 0 & 0 \end{pmatrix}$$

Trigonal: upper (4)

$$\begin{pmatrix} 111 & 0 & 0 & 0 & -111 & -231 & 0 & 123 & 0 \\ 0 & -111 & 231 & -111 & 0 & 0 & -123 & 0 & 0 \\ 0 & 321 & 0 & -321 & 0 & 0 & 0 & 0 & 0 \end{pmatrix}$$

Trigonal: lower (9)

$$\begin{pmatrix} 111 & 121 & 131 & -121 & -111 & -231 & 113 & 123 & 0 \\ 121 & -111 & 231 & -111 & -121 & 131 & -123 & 113 & 0 \\ 311 & 321 & 0 & -321 & 311 & 0 & 0 & 0 & 333 \end{pmatrix}$$

Tetragonal and hexagonal: upper (3)

$$\begin{pmatrix} 0 & 0 & 0 & 0 & 0 & -231 & 0 & 123 & 0 \\ 0 & 0 & 231 & 0 & 0 & 0 & -123 & 0 & 0 \\ 0 & 321 & 0 & -321 & 0 & 0 & 0 & 0 & 0 \end{pmatrix}$$

Tetragonal and hexagonal: lower (7)

$$\begin{pmatrix} 0 & 0 & 131 & 0 & 0 & -231 & 113 & 123 & 0 \\ 0 & 0 & 231 & 0 & 0 & 131 & -123 & 113 & 0 \\ 311 & 321 & 0 & -321 & 311 & 0 & 0 & 0 & 333 \end{pmatrix}$$

Cubic: upper (1)

$$\begin{pmatrix} 0 & 0 & 0 & 0 & 0 & -231 & 0 & 231 & 0 \\ 0 & 0 & 231 & 0 & 0 & 0 & -231 & 0 & 0 \\ 0 & -231 & 0 & 231 & 0 & 0 & 0 & 0 & 0 \end{pmatrix}$$

Cubic: lower (2)

$$\begin{pmatrix} 0 & 0 & 0 & 0 & 0 & 321 & 0 & 231 & 0 \\ 0 & 0 & 231 & 0 & 0 & 0 & 321 & 0 & 0 \\ 0 & 321 & 0 & 231 & 0 & 0 & 0 & 0 & 0 \end{pmatrix}$$

As already mentioned, there are no properties of $T(3)$ type discussed in Chapters 1 or 2, but higher-order effects of this type do exist, and could conceivably be of interest in some experimental situations.

Problems

7–1. Work out the coefficients of the partly symmetric matter tensor $T(3)$ for the point group D_{3h}.

7–2. Do the same for the following groups involving $\tilde{E}$-type irreps: S_4, C_{3h}, C_4. In each case, compare your results with Table 7–1.

7–3. Show why the engineering piezoelectric coefficients d^{e}_{ij} may be obtained from the K_{ij} coefficients of Table 7–1 simply by changing all factors of $\sqrt{2}$ into factors of 2.

7–4. Show how the engineering electro-optic coefficients r^{e}_{ij} are obtained from the K_{ij} coefficients of Table 7–1.

7–5. Work out the two-index and three-index coefficients for a $T(3)^{ax}$ tensor for the lower trigonal and monoclinic crystal classes, and compare your results with Tables 7–2 and 7–3.

7–6. Interpret the meaning of the non-zero K coefficients of the Nernst effect (see Eq. (2–28)) for the case of an upper hexagonal or tetragonal crystal, as given in Table 7–3. What is the physical meaning of the additional coefficients that appear for crystals of the lower hexagonal symmetry?

7–7. Do the same analysis for the Ettinghausen effect (the converse of the Nernst effect). (See Eq. (2–29).)

8

Special magnetic properties

In Chapter 5, it was shown that time reversal or magnetic symmetry becomes important only when we deal with 'special magnetic properties', defined by the following two criteria:

1. The matter tensor **K** is a *c*-tensor, due to the fact that *either* **X** *or* **Y** is a *c*-tensor, *and*
2. the property in question is not a transport property (involving an increase in entropy).

Such properties are given an asterisk in Table 1–1. In dealing with these properties we must take cognizance of the magnetic symmetry of the crystals in which they are observed, that is, of the 90 magnetic classes, as distinct from the 32 conventional crystal classes which sufficed for the study of all other properties.

Since the principal special magnetic properties of interest do not involve tensors of rank higher than 3 (see Table 1–1), and such ranks have already been covered for the conventional properties in Chapters 6 and 7, a diversion to special magnetic properties at this point seems appropriate. We then resume the main thrust of the book with Chapter 9, that is, continuing to consider higher tensor ranks.

As shown in Chapter 5, if **K** represents a special magnetic property, it must be identically zero for non-magnetic (i.e. diamagnetic or paramagnetic) crystals, as well as for antiferromagnetic crystals belonging to type-II groups. Therefore, the special magnetic properties which we consider in this chapter only exist for crystals that possess magnetic symmetry, namely, those of types I and III (see Section 5–1). In treating these properties we will proceed, in the usual manner, to consider successively increasing tensor ranks, from 1 to 2 to 3, drawing upon material in Chapters 6 and 7 where appropriate.

8–1 *c*-tensors of rank 1: the magnetocaloric effect

The special magnetic properties of rank 1 are relations between a scalar and an axial vector, so that **K** is a tensor of type $T(1)^{ax}$. Depending on whether **X** or **Y** is the scalar determines whether **K** is a column or a row of three components. The example given in Table 1–1 is the magnetocaloric effect or its converse, the pyromagnetic effect. For the pyromagnetic effect, **K** is a column vector while for the magnetocaloric effect, the same three components appear as a row vector.

To proceed to identify which of the 90 magnetic groups can show these effects, we require that the rep $\Gamma_{\mathbf{R}}$ for a type-I group or the complementary rep $\Gamma^{c}_{\mathbf{R}}$ of a type-III group possess invariant components (belonging to the A_1 irrep). But the decomposition of these reps into irreps is already given in Table 5–1 for all 90 magnetic groups. We leave it to the reader to show that all groups containing the complementary operation $\underline{i}$ are excluded. In addition, many others are also excluded, leaving only 31 of the 90 magnetic classes for which **K** is not identically zero. With the aid of Table 5–1 it is trivial to collect these components and to obtain the results of Table 8–1 where **K** is shown in the form of a row. Note that only in the triclinic classes do all three components appear. For three of the monoclinic classes only two components appear, while, for the remaining classes there is only a single component. For all cubic groups $\mathbf{K} \equiv 0$.

It is interesting to note the variability of the **K** vector for different magnetic groups derived from the same classical group. Thus, for example, for the magnetic groups derived from C_{2h}, the group C_{2h} itself has only a K_3 component, $C_{2h}{:}C_i$ has only K_1 and K_2, while for variants $C_{2h}{:}C_s$ and $C_{2h}{:}C_2$, $\mathbf{K} \equiv 0$. These examples illustrate the role of magnetic symmetry quite strikingly.

It is important to note that the magnetic classes that exhibit the property of pyromagnetism are precisely those that can exhibit a spontaneous magnetization (i.e. magnetization in the absence of a magnetic field). Such crystals are the ferromagnetics and ferrimagnetics discussed in Chapter 5. The two properties are intimately interlinked. (See Problem 8–1.)

Among the specific examples of crystals that fall into these categories are several that are basically antiferromagnetic crystals but do possess a weak spontaneous magnetization due to a slight tilting of the spins away from exact antiparallel alignment. Such tilting is only allowable if the crystal belongs to one of the magnetic classes of Table 8–1. Examples are NiF_2, α-Fe_2O_3 (hematite) and $MnCO_3$, among others.

A good illustration comes from the case of NiF_2. The atomic arrangement in this crystal is that of the well-known rutile structure which is tetragonal (D_{4h} or $4/mmm$ symmetry), as shown in the atomic arrangement of Fig. 8–1.* Other fluorides that have this same atomic structure are MgF_2, MnF_2, and CoF_2. But when spins are included they are all totally different. MgF_2 is non-magnetic and possesses no spin structure, while in MnF_2 and CoF_2 the spins are alternately parallel and antiparallel to the four-fold c-axis, as shown in Fig. 8–1(a). Here the appropriate magnetic group is D_{4h}:D_{2h}. This group does not allow for spontaneous magnetization (i.e. it does not appear in Table 8–1) and, accordingly, the crystal is a perfect antiferromagnet. Any tilting of the spins would destroy the four-fold axis and, therefore, change the crystal symmetry. But for NiF_2, the antiferromagnetic spins are parallel and antiparallel to the x-axis, as shown in Fig. 8–1(b). The appropriate magnetic group is then D_{2h}:C_{2h} (an orthorhombic group) for which spontaneous magnetization is possible along the preferred axis of the C_{2h} subgroup, which here is the y-axis. Thus, weak ferromagnetism (i.e. a net magnetic moment) is obtained by a rotation of the spins slightly away from parallelism to x toward direction Oy, without changing the D_{2h}:C_{2h} symmetry of the crystal as shown in Fig. 8–2. In the cases of α-Fe_2O_3 and $MnCO_3$, the same type of weak ferromagnetism develops but with a different crystal structure.

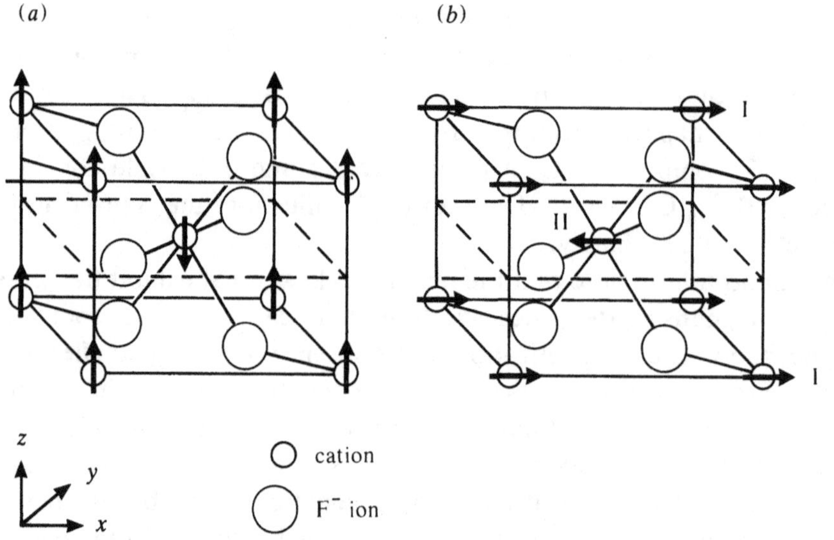

Fig. 8–1. Comparison of magnetic structures of various crystals possessing the rutile structure: (*a*) MnF_2 or CoF_2; (*b*) NiF_2.

* The reader should realize that some of the symmetry operations of the rutile structure involve translations to produce screw axes and glide planes. For example, the C_4 operation is not a pure rotation but a screw rotation with translation by (1/2, 1/2, 1/2).

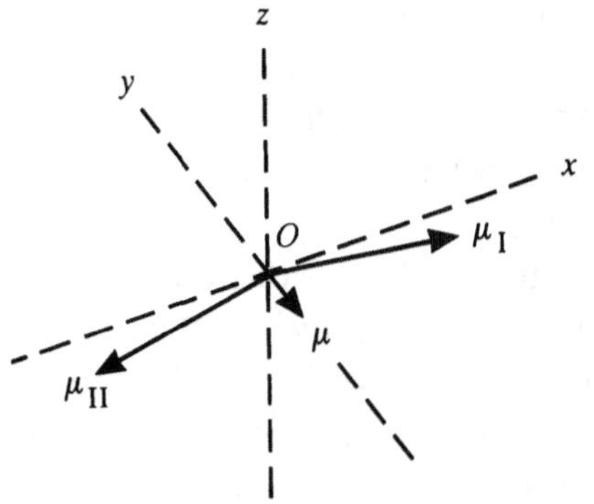

Fig. 8–2. Showing how non-collinearity of spins in NiF_2 leads to a net magnetic moment, μ, in the OY direction. [I and II refer to the two cationic spins of Fig. 8–1(*b*).]

Table 8–1. $T(1)^{ax}$ *tensors; special magnetic*

Form of **K**	Magnetic groups
(K_1, K_2, K_3)	C_1, C_i
$(K_1, K_2, 0)$	C_s:C_1, C_2:C_1, C_{2h}:C_i
$(0, K_2, 0)$	C_{2v}:C_s
$(0, 0, K_3)$	C_s, C_2, C_{2h}, C_{2v}:C_2, D_2:C_2, D_{2h}:C_{2h}, C_4, S_4, C_{4h}, C_{4v}:C_4, C_{4v}:S_4, D_4:C_4, D_{4h}:C_{4h}, C_3, S_6, D_3:C_3, C_{3v}:C_3, D_{3d}:S_6, C_6, C_{3h}, C_{6h}, D_6:C_6, C_{6v}:C_6, D_{3h}:C_{3h}, D_{6h}:C_{6h}

For all other groups $\mathbf{K} = 0$.

Apart from the magnetocaloric effect and its converse, the considerations of this section may also be applied to a restricted form of the piezomagnetic effect, namely, the effect of hydrostatic pressure, P, in producing magnetization (or magnetic induction). Since P is invariant, like ΔT, the symmetry considerations are identical to the pyromagnetic effect, and Table 8–1 may also be regarded as a listing of **K** vectors representing this effect. The converse effect is that of a magnetic field giving rise to dilatational strain. The full treatment of the piezomagnetic effect, however, will be given later in this chapter.

8–2 *c*-tensors of rank 2: the magnetoelectric effect

The next higher special magnetic property is one of type $T(2)^{ax}$, for example the magnetoelectric polarizability λ_{ij}. This is a *c*-tensor which relates a magnetic field – the axial vector $\mathbf{X}$ – to an electric displacement vector – the polar vector $\mathbf{Y}$. Its converse λ'_{ij} relates an electric field to a magnetic induction and is simply the transpose of the tensor $\boldsymbol{\lambda}$ (see Eq. (1–15)). To be specific, we here select the direct effect, so that $\mathbf{X}$ is a $T(1)^{ax}$ and $\mathbf{Y}$ is a T(1).

As for all special magnetic effects, this effect vanishes for non-magnetic crystals. It also vanishes for many of the 90 magnetic groups. The procedure for obtaining the form of $\mathbf{K}$ (i.e. $\boldsymbol{\lambda}$) is to look for common irreps between Γ_{xyz} and $\Gamma_{\mathbf{R}}$ (in the case of type-I) or $\Gamma^{c}_{\mathbf{R}}$ (in the case of type-III magnetic groups). It is immediately clear that $\mathbf{K} \equiv 0$ for the 21 magnetic classes containing the inversion operation i. (Why?) However, $\mathbf{K}$ also vanishes in another 11 classes where there are no common irreps, so that only 58 classes are potentially capable of exhibiting this effect.

By way of illustration, consider the magnetic group D_{2d}:C_{2v}. The corresponding classical group is D_{2d}, for which $\Gamma_{xyz} = B_2 + E$ (see the S-C-T table) with the orientation of coordinates for irrep E given by (x, y), and $\Gamma_{\mathbf{R}} = A_2 + E$ with E-orientation $(R_x, -R_y)$. Time reversal has no effect on Γ_{xyz} but changes $\Gamma_{\mathbf{R}}$ into $\Gamma^{c}_{\mathbf{R}} = B_1 + E$ with E-orientation now given by $(R_y, -R_x)$ (see Table 5–1). Thus, in D_{2d}:C_{2v} there is no common irrep for coordinates z and R_z but the remaining coordinates belong to irrep E, in which $(R_y, -R_x)$ is oriented similarly to (x, y). Accordingly, the form of $\bar{\mathbf{K}}$ is

$$\bar{\mathbf{K}} = \begin{pmatrix} 0 & K_E & 0 \\ -K_E & 0 & 0 \\ 0 & 0 & 0 \end{pmatrix} \tag{8–1}$$

with only a single constant appearing, as shown.

For magnetic group C_4, we have $\Gamma_{xyz} = A + \tilde{E}$ and $\Gamma_{\mathbf{R}} = A + \tilde{E}$ with $x \mp iy$ matching $R_x \mp iR_y$. Then, from Eqs. (4–22, 4–23) we obtain

$$\mathbf{K} = \bar{\mathbf{K}} = \begin{pmatrix} K^{re}_{\tilde{E}} & K^{im}_{\tilde{E}} & 0 \\ -K^{im}_{\tilde{E}} & K^{re}_{\tilde{E}} & 0 \\ 0 & 0 & K_A \end{pmatrix} \tag{8–2}$$

Thus $\mathbf{K}$ takes the form shown in Table 8–2 with

$$\left.\begin{aligned} K_{11} &= K^{re}_{\tilde{E}} \\ K_{12} &= K^{im}_{\tilde{E}} \\ K_{33} &= K_A \end{aligned}\right\} \tag{8–3}$$

For magnetic group C_4:C_2, on the other hand, we must compare $\Gamma_{xyz} = A + \tilde{E}$ to $\Gamma^{c}_{\mathbf{R}} = B + \tilde{E}$ with $x \mp iy$ now matched by $R_x \pm iR_y$. Thus, we now find $K_{33} = 0$ and

$$\mathbf{K} = \bar{\mathbf{K}} = \begin{pmatrix} K_{\tilde{\mathrm{E}}}^{\mathrm{re}} & K_{\tilde{\mathrm{E}}}^{\mathrm{im}} & 0 \\ K_{\tilde{\mathrm{E}}}^{\mathrm{im}} & -K_{\tilde{\mathrm{E}}}^{\mathrm{re}} & 0 \\ 0 & 0 & 0 \end{pmatrix} \tag{8–4}$$

In this way, we obtain Table 8–2 for all the 58 classes that can exhibit this effect. Clearly, the magnetic groups of type I (i.e. those without complementary operations) give the same results as those obtained in Chapter 6 for $T(2)^{\mathrm{ax}}$ tensors of the same classical group (Table 6–3).

An interesting approach is one in which, instead of starting from the magnetic symmetry and analyzing the properties, one does the reverse. An X-ray structural study of Cr_2O_3 shows that it possesses the classical symmetry D_{3d}. If this were also its magnetic structure in its antiferromagnetic state, the magnetoelectric effect would be forbidden. Instead, one finds two constants: $\lambda_{11} = \lambda_{22}$ and λ_{33}. Comparison with Table 8–2 then shows that the magnetic symmetry of the antiferromagnetic state of Cr_2O_3 must be D_{3d}:D_3.

8–3 *c*-tensors of rank 3: the piezomagnetic effect

The interaction between a $T_S(2)$ quantity (e.g. stress or strain) and a $T(1)^{\mathrm{ax}}$ magnetic quantity (e.g. magnetic field or induction) produces a $T(3)^{\mathrm{ax}}$ *c*-tensor with intrinsic two-index symmetry. Examples are the piezomagnetic effect Q_{ijk} and its converse Q'_{ijk} which are related by Eq. (1–17). Here we take **X** as the $T_S(2)$ quantity and **Y** as $T(1)^{\mathrm{ax}}$ (as for Q_{ijk}) and use the Weyl normalization to convert the $T_S(2)$ quantity into a six-vector (Eq. (4–4)). Then **K** can be regarded as a 3×6 matrix, similarly to the piezoelectric effect in Chapter 6. However, since the present tensor **K** is a special magnetic property, it is again prohibited for all non-magnetic crystals.

For the field **X**, $\Gamma = \Gamma_\alpha$, while for the response **Y**, $\Gamma = \Gamma_{\mathbf{R}}$ for type-I and $\Gamma_{\mathbf{R}}^{\mathrm{c}}$ for type-III magnetic groups. We therefore seek common irreps of Γ_α and either $\Gamma_{\mathbf{R}}$ or $\Gamma_{\mathbf{R}}^{\mathrm{c}}$. It is immediately apparent that for groups containing the complementary operation $\underline{\mathrm{i}}$, all Γ_α irreps are of the g-type while all $\Gamma_{\mathbf{R}}^{\mathrm{c}}$ irreps are of the u-type. Therefore, this effect is excluded for the 21 magnetic classes containing $\underline{\mathrm{i}}$. In order to obtain the specific form of **K** when it does not vanish identically, we need only compare $\Gamma_{\mathbf{R}}$ or $\Gamma_{\mathbf{R}}^{\mathrm{c}}$ as given in Table 5–1 with Γ_α given in the S-C-T table. We illustrate with some examples.

Consider, first, the classical group C_{3v} for which there are two magnetic groups: C_{3v} and C_{3v}:C_3. For C_{3v}, $\Gamma_{\mathbf{R}} = A_2 + E$ with R_z belonging to irrep A_2 and $(R_x, -R_y)$ to E with the same orientation as (y, x). From the

Table 8–2. T$(2)^{ax}$ *tensors; special magnetic*

Magnetic groups C_1, C_i $\begin{pmatrix} K_{11} & K_{12} & K_{13} \\ K_{21} & K_{22} & K_{23} \\ K_{31} & K_{32} & K_{33} \end{pmatrix}$	C_2, $C_s{:}C_1$, $C_{2h}{:}C_2$ $\begin{pmatrix} K_{11} & K_{12} & 0 \\ K_{21} & K_{22} & 0 \\ 0 & 0 & K_{33} \end{pmatrix}$	C_s, $C_2{:}C_1$, $C_{2h}{:}C_s$ $\begin{pmatrix} 0 & 0 & K_{13} \\ 0 & 0 & K_{23} \\ K_{31} & K_{32} & 0 \end{pmatrix}$
C_{2v}, $D_{2h}{:}C_{2v}$, $D_2{:}C_2$ $\begin{pmatrix} 0 & K_{12} & 0 \\ K_{21} & 0 & 0 \\ 0 & 0 & 0 \end{pmatrix}$	$C_{2v}{:}C_2$, D_2, $D_{2h}{:}D_2$ $\begin{pmatrix} K_{11} & 0 & 0 \\ 0 & K_{22} & 0 \\ 0 & 0 & K_{33} \end{pmatrix}$	$C_{2v}{:}C_s$ $\begin{pmatrix} 0 & 0 & 0 \\ 0 & 0 & K_{23} \\ 0 & K_{32} & 0 \end{pmatrix}$
C_4, $C_4{:}C_2$, $C_{4h}{:}C_4$, C_3, $S_6{:}C_3$, C_6, $C_{3h}{:}C_3$, $C_{6h}{:}C_6$ $\begin{pmatrix} K_{11} & K_{12} & 0 \\ -K_{12} & K_{11} & 0 \\ 0 & 0 & K_{33} \end{pmatrix}$	D_4, $C_{4v}{:}C_4$, $D_{2d}{:}D_2$, D_3, $D_{4h}{:}D_4$, $D_{3d}{:}D_3$, $C_{3v}{:}C_3$, $C_{6v}{:}C_6$, $D_{3h}{:}D_3$, D_6, $D_{6h}{:}D_6$ $\begin{pmatrix} K_{11} & 0 & 0 \\ 0 & K_{11} & 0 \\ 0 & 0 & K_{33} \end{pmatrix}$	$C_4{:}C_2$, S_4, $C_{4h}{:}S_4$ $\begin{pmatrix} K_{11} & K_{12} & 0 \\ K_{12} & -K_{11} & 0 \\ 0 & 0 & 0 \end{pmatrix}$
$C_{4v}{:}C_{2v}$, $D_{2d}{:}S_4$ $\begin{pmatrix} 0 & K_{12} & 0 \\ K_{12} & 0 & 0 \\ 0 & 0 & 0 \end{pmatrix}$	$D_4{:}D_2$, D_{2d}, $D_{4h}{:}D_{2d}$ $\begin{pmatrix} K_{11} & 0 & 0 \\ 0 & -K_{11} & 0 \\ 0 & 0 & 0 \end{pmatrix}$	C_{4v}, $D_{2d}{:}C_{2v}$, $D_4{:}C_4$, $D_{4h}{:}C_{4v}$, C_{3v}, $D_3{:}C_3$, $D_{3d}{:}C_{3v}$, $D_{3h}{:}C_{3v}$, C_{6v}, $D_6{:}C_6$, $D_{6h}{:}C_{6v}$ $\begin{pmatrix} 0 & K_{12} & 0 \\ -K_{12} & 0 & 0 \\ 0 & 0 & 0 \end{pmatrix}$
T, $T_h{:}T$, $T_d{:}T$, O, $O_h{:}O$ $\begin{pmatrix} K_{11} & 0 & 0 \\ 0 & K_{11} & 0 \\ 0 & 0 & K_{11} \end{pmatrix}$		

S-C-T table, we find that $\Gamma_\alpha = 2A_1 + 2E$, with $\alpha_1 + \alpha_2$ and α_3 belonging to A_1, and that (α_4, α_5) as well as $(\alpha_1 - \alpha_2, \alpha_6)$ belong to E, with proper orientation. Thus there is no common irrep between the α-components and R_z, and so the bottom row of our **K** matrix is identically zero. Taking the symmetry coordinates of **X** in the order: $(X_1 + X_2)/\sqrt{2}$, $(X_1 - X_2)/\sqrt{2}$, X_3, X_4, X_5 and X_6, the matrix $\bar{\mathbf{K}}$ becomes

$$\bar{\mathbf{K}} = \begin{pmatrix} 0 & K_{E,2} & 0 & K_{E,1} & 0 & 0 \\ 0 & 0 & 0 & 0 & -K_{E,1} & -K_{E,2} \\ 0 & 0 & 0 & 0 & 0 & 0 \end{pmatrix} \tag{8–5}$$

from which we may return to the original coordinates to obtain the form of the **K** matrix in Table 8–3 in which

$$\left.\begin{aligned} K_{11} &= K_{E,2}/\sqrt{2} \\ K_{14} &= K_{E,1} \end{aligned}\right\} \tag{8–6}$$

The other magnetic group C_{3v}:C_3, on the other hand, has the same Γ_α, but now $\Gamma^c_{\mathbf{R}} = A_1 + E$, with R_z belonging to A_1 and (R_y, R_x) to E. We thus now have non-zero components K_{3j}. The matrix $\bar{\mathbf{K}}$ is then

$$\bar{\mathbf{K}} = \begin{pmatrix} 0 & 0 & 0 & 0 & K_{E,1} & K_{E,2} \\ 0 & K_{E,2} & 0 & K_{E,1} & 0 & 0 \\ K_{A,1} & 0 & K_{A,2} & 0 & 0 & 0 \end{pmatrix} \tag{8–7}$$

Returning to the original coordinates, we obtain **K** in the form given in Table 8–3, with

$$\left.\begin{aligned} K_{15} &= K_{E,1}, \qquad & K_{22} &= -K_{E,2}/\sqrt{2} \\ K_{31} &= K_{A,1}/\sqrt{2}, \qquad & K_{33} &= K_{A,2} \end{aligned}\right\} \tag{8–8}$$

We further illustrate by considering magnetic groups that involve the complex-pair $\tilde{E}$ irreps, as in the groups stemming from C_{6h}. We begin with C_{6h} itself, for which $\Gamma_{\mathbf{R}} = A_g + \tilde{E}_{1g}$ with R_z in A_g and $R_x \mp iR_y$ in $\tilde{E}_{1g}$, while $\Gamma_\alpha = 2A_g + \tilde{E}_{1g} + \tilde{E}_{2g}$. Here $\alpha_1 + \alpha_2$ and α_3 belong to A_g, and $\alpha_4 \pm i\alpha_5$ to $\tilde{E}_{1g}$, while $\alpha_1 - \alpha_2 \mp i\sqrt{2}\alpha_6$ belongs to $\tilde{E}_{2g}$. Thus, there are two constants related to irrep A_g due to the repeat, and two $\tilde{E}_{1g}$ constants of the form given in Eq. (4–25). Using the same symmetry coordinates of **X** as in the case of C_{3v}, we then find

$$\bar{\mathbf{K}} = \begin{pmatrix} 0 & 0 & 0 & \tilde{E}^{re}_{1g} & \tilde{E}^{im}_{1g} & 0 \\ 0 & 0 & 0 & \tilde{E}^{im}_{1g} & -\tilde{E}^{re}_{1g} & 0 \\ A_g,1 & 0 & A_g,2 & 0 & 0 & 0 \end{pmatrix} \tag{8–9}$$

where the same simplifying convention used in Eq. (7–33) is again used, namely, showing only the sign, the subscript and the superscript of each $\bar{\mathbf{K}}$

component that is not zero. This result leads to the form of the **K** matrix given in Table 8–3, with

$$\left.\begin{array}{ll} K_{31} = K_{\mathrm{A},1}/\sqrt{2}, & K_{33} = K_{\mathrm{A},2} \\ K_{14} = K_{\tilde{\underline{\mathrm{E}}}}^{\mathrm{re}} \quad \text{and} & K_{15} = K_{\tilde{\underline{\mathrm{E}}}}^{\mathrm{im}} \end{array}\right\} \qquad (8\text{–}10)$$

For the magnetic groups $\mathrm{C_{6h}}$:$\mathrm{C_6}$ and $\mathrm{C_{6h}}$:$\mathrm{C_{3h}}$, both of which have $\underline{\mathrm{i}}$ as an element, only irreps of u-type are obtained for $\Gamma_{\mathbf{R}}^{\mathrm{c}}$, so that **K** is identically zero. This leaves the magnetic group $\mathrm{C_{6h}}$:$\mathrm{S_6}$ for which $\Gamma_{\mathbf{R}}^{\mathrm{c}} = \mathrm{B_g} + \tilde{\mathrm{E}}_{2\mathrm{g}}$, with R_z belonging to $\mathrm{B_g}$ and $R_x \pm \mathrm{i}R_y$ to $\tilde{\mathrm{E}}_{2\mathrm{g}}$. The rep Γ_α, as given above, does not contain B , but has $\alpha \quad - \alpha \ \mp \mathrm{i}\sqrt{2}\alpha$ in $\tilde{\mathrm{E}}$, leading to a form for $\bar{\mathbf{K}}$:

$$\bar{\mathbf{K}} = \begin{pmatrix} 0 & \tilde{\mathrm{E}}_{2\mathrm{g}}^{\mathrm{re}} & 0 & 0 & 0 & \tilde{\mathrm{E}}_{2\mathrm{g}}^{\mathrm{im}} \\ 0 & \tilde{\mathrm{E}}_{2\mathrm{g}}^{\mathrm{im}} & 0 & 0 & 0 & -\tilde{\mathrm{E}}_{2\mathrm{g}}^{\mathrm{re}} \\ 0 & 0 & 0 & 0 & 0 & 0 \end{pmatrix} \qquad (8\text{–}11)$$

This result leads to the form for **K** given in Table 8–3, with

$$K_{11} = \tilde{\mathrm{E}}_{2\mathrm{g}}^{\mathrm{re}}/\sqrt{2}, \quad K_{22} = -\tilde{\mathrm{E}}_{2\mathrm{g}}^{\mathrm{im}}/\sqrt{2} \qquad (8\text{–}12)$$

In a similar way, the reader may obtain the forms of **K** given in Table 8–3 for other magnetic groups. (See the problems.)

It is interesting to note that for all three cases of special magnetic properties, $T(1)^{\mathrm{ax}}$, $T(2)^{\mathrm{ax}}$ and $T(3)^{\mathrm{ax}}$, covered, respectively, by Tables 8–1, 8–2 and 8–3, the general rule is that no two magnetic variants of a given classical group have the same matrix of coefficients. This rule is not obeyed in tables given in the literature, however (e.g. see Birss, 1964), where different results from those in Table 8–3 are obtained for magnetic classes $\mathrm{D_{2d}}$:$\mathrm{C_{2v}}$, $\mathrm{D_{3h}}$:$\mathrm{D_3}$ and $\mathrm{C_{2v}}$:$\mathrm{C_s}$ as a consequence of a different choice of axes from the ones we have made in Table 5–1. (See, for example, Problem 8–4.)

As already mentioned, the principal example of this type of partly symmetric $T(3)^{\mathrm{ax}}$ matter tensor is that of the piezomagnetic effect Q_{ijk}. In this case **X** is the stress tensor and **Y** the magnetic induction **B**. As in the case of the piezoelectric effect, it is customary in dealing with piezomagnetism to change the three-index tensor to a two-index engineering matrix Q_{ij}^{e} in the same way as Eqs. (7–3). The resulting table of two-index coefficients is then the same as that in Table 8–3, except that all factors $\sqrt{2}$ in the table become factors of 2. Also, for the piezomagnetic coefficients, we may connect the two-index K_{ij}'s of Table 8–3 to three-index tensor quantities using Eq. (7–2). In both these index conversions, there is a close similarity to the case of the piezoelectric effect, for obvious reasons.

Despite the large number of magnetic classes for which the piezomagnetic effect is permissible, it has apparently only been studied experimentally in the antiferromagnetic crystals MnF_2, CoF_2 and $FeCO_3$ (Shuvalov, 1988). We note that MnF_2 and CoF_2 belong to the magnetic class D_{4h}:D_{2h} (see Fig. 8–1(a)), and so, from Table 8–3, that there are only two Q_{ij} coefficients, namely, Q_{14} and Q_{36}. The values of these coefficients for CoF_2 at 20.4 K, in units of gauss $kg^{-1}\,cm^2$, were found to be $Q_{14} = 2.1 \times 10^{-3}$ and $Q_{36} = 0.8 \times 10^{-3}$. These same coefficients are ~100 times smaller for MnF_2.

Aside from crystals as piezomagnetics, a magnetostrictive metal or ceramic (polycrystalline) can be given piezomagnetic properties similar to a hexagonal (C_{6v}) crystal by poling it with a magnetic field. Such a material can be useful for transducers (analogous to piezoelectric transducers) to generate longitudinal, torsional and shear modes of motion (Mason, 1966).

8–4 Symmetric *c*-tensors of rank 3: higher-order magnetic permeability

Another special magnetic property of rank 3 is obtained by expanding the B–H relation, Eq. (1–11), to higher terms, so that the permeability μ becomes a function of the magnetic field:

$$\mu_{ij} = \mu_{ij}^0 + P_{ijk}H_k \tag{8–13}$$

The tensor P_{ijk} is readily seen to be of third rank and axial. In addition, by setting up a magnetic thermodynamic potential (in a manner analogous to Eq. (1–6)):

$$\Phi \propto P_{ijk}H_iH_jH_k \tag{8–14}$$

it can be seen that P_{ijk} possesses intrinsic symmetry in all three coefficients: $i \leftrightarrow j \leftrightarrow k$. Thus **P** is a $T_S(3)^{ax}$ tensor and is also a special magnetic property. Because of the additional intrinsic symmetry, the P_{ijk} coefficients will contain fewer independent values than the K_{ijk} of the preceding section. It can, however, be obtained from the tensors of Table 8–3 by setting appropriate coefficients equal. Thus, the full symmetry of the three coefficients means that in the two-index notation of Table 8–3 (utilizing Eq. (7–2)):

$$\left.\begin{array}{lll} K_{12} = K_{26}/\sqrt{2}; & K_{13} = K_{35}/\sqrt{2}; & K_{31} = K_{15}/\sqrt{2} \\ K_{23} = K_{34}/\sqrt{2}; & K_{32} = K_{24}/\sqrt{2}; & K_{21} = K_{16}/\sqrt{2} \\ K_{14} = K_{25} = K_{36} & & \end{array}\right\} \tag{8–15}$$

Table 8–3. $T(3)^{ax}$ *tensors; special magnetic (two-index form)*

Magnetic groups C_1, C_i

$$\begin{pmatrix} K_{11} & K_{12} & K_{13} & K_{14} & K_{15} & K_{16} \\ K_{21} & K_{22} & K_{23} & K_{24} & K_{25} & K_{26} \\ K_{31} & K_{32} & K_{33} & K_{34} & K_{35} & K_{36} \end{pmatrix}$$

C_s, C_2, C_{2h}

$$\begin{pmatrix} 0 & 0 & 0 & K_{14} & K_{15} & 0 \\ 0 & 0 & 0 & K_{24} & K_{25} & 0 \\ K_{31} & K_{32} & K_{33} & 0 & 0 & K_{36} \end{pmatrix}$$

$C_s{:}C_1$, $C_2{:}C_1$, $C_{2h}{:}C_i$

$$\begin{pmatrix} K_{11} & K_{12} & K_{13} & 0 & 0 & K_{16} \\ K_{21} & K_{22} & K_{23} & 0 & 0 & K_{26} \\ 0 & 0 & 0 & K_{34} & K_{35} & 0 \end{pmatrix}$$

$C_{2v}{:}C_2$, $D_2{:}C_2$, $D_{2h}{:}C_{2h}$

$$\begin{pmatrix} 0 & 0 & 0 & 0 & K_{15} & 0 \\ 0 & 0 & 0 & K_{24} & 0 & 0 \\ K_{31} & K_{32} & K_{33} & 0 & 0 & 0 \end{pmatrix}$$

$C_{2v}{:}C_s$

$$\begin{pmatrix} 0 & 0 & 0 & 0 & 0 & K_{16} \\ K_{21} & K_{22} & K_{33} & 0 & 0 & 0 \\ 0 & 0 & 0 & K_{34} & 0 & 0 \end{pmatrix}$$

C_3, S_6

$$\begin{pmatrix} K_{11} & -K_{11} & 0 & K_{14} & K_{15} & -\sqrt{2}K_{22} \\ -K_{22} & K_{22} & 0 & K_{15} & -K_{14} & -\sqrt{2}K_{11} \\ K_{31} & K_{31} & K_{33} & 0 & 0 & 0 \end{pmatrix}$$

$D_3{:}C_3$, $C_{3v}{:}C_3$, $D_{3d}{:}S_6$

$$\begin{pmatrix} 0 & 0 & 0 & 0 & K_{15} & -\sqrt{2}K_{22} \\ -K_{22} & K_{22} & 0 & K_{15} & 0 & 0 \\ K_{31} & K_{31} & K_{33} & 0 & 0 & 0 \end{pmatrix}$$

$C_4{:}C_2$, $S_4{:}C_2$, $C_{4h}{:}C_{2h}$

$$\begin{pmatrix} 0 & 0 & 0 & K_{14} & K_{15} & 0 \\ 0 & 0 & 0 & -K_{15} & K_{14} & 0 \\ K_{31} & -K_{31} & 0 & 0 & 0 & K_{36} \end{pmatrix}$$

$D_{2d}{:}C_{2v}$

$$\begin{pmatrix} 0 & 0 & 0 & 0 & K_{15} & 0 \\ 0 & 0 & 0 & -K_{15} & 0 & 0 \\ K_{31} & -K_{31} & 0 & 0 & 0 & 0 \end{pmatrix}$$

C_{4h}, C_4, S_4, C_6, C_{3h}, C_{6h}

$$\begin{pmatrix} 0 & 0 & 0 & K_{14} & K_{15} & 0 \\ 0 & 0 & 0 & K_{15} & -K_{14} & 0 \\ K_{31} & K_{31} & K_{33} & 0 & 0 & 0 \end{pmatrix}$$

Table 8–3. *(cont.)*

C_{2v}, D_2, D_{2h}

$$\begin{pmatrix} 0 & 0 & 0 & K_{14} & 0 & 0 \\ 0 & 0 & 0 & 0 & K_{25} & 0 \\ 0 & 0 & 0 & 0 & 0 & K_{36} \end{pmatrix}$$

$C_{4v}{:}C_{2v}$, $D_4{:}D_2$, $D_{2d}{:}D_2$, $D_{4h}{:}D_{2h}$

$$\begin{pmatrix} 0 & 0 & 0 & K_{14} & 0 & 0 \\ 0 & 0 & 0 & 0 & K_{14} & 0 \\ 0 & 0 & 0 & 0 & 0 & K_{36} \end{pmatrix}$$

$C_{6h}{:}S_6$, $C_6{:}C_3$, $C_{3h}{:}C_3$

$$\begin{pmatrix} K_{11} & -K_{11} & 0 & 0 & 0 & -\sqrt{2}K_{22} \\ -K_{22} & K_{22} & 0 & 0 & 0 & -\sqrt{2}K_{11} \\ 0 & 0 & 0 & 0 & 0 & 0 \end{pmatrix}$$

$D_4{:}C_4$, $C_{4v}{:}C_4$, $D_{2d}{:}S_4$, $D_{4h}{:}C_{4h}$, $D_6{:}C_6$, $C_{6v}{:}C_6$, $D_{3h}{:}C_{3h}$, $D_{6h}{:}C_{6h}$

$$\begin{pmatrix} 0 & 0 & 0 & 0 & K_{15} & 0 \\ 0 & 0 & 0 & K_{15} & 0 & 0 \\ K_{31} & K_{31} & K_{33} & 0 & 0 & 0 \end{pmatrix}$$

D_3, C_{3v}, D_{3d}

$$\begin{pmatrix} K_{11} & -K_{11} & 0 & K_{14} & 0 & 0 \\ 0 & 0 & 0 & 0 & -K_{14} & -\sqrt{2}K_{11} \\ 0 & 0 & 0 & 0 & 0 & 0 \end{pmatrix}$$

$D_6{:}D_3$, $D_{6h}{:}D_{3d}$, $D_{3h}{:}D_{3v}$, $C_{6v}{:}C_{3v}$

$$\begin{pmatrix} K_{11} & -K_{11} & 0 & 0 & 0 & 0 \\ 0 & 0 & 0 & 0 & 0 & -\sqrt{2}K_{11} \\ 0 & 0 & 0 & 0 & 0 & 0 \end{pmatrix}$$

$D_{3h}{:}D_3$

$$\begin{pmatrix} 0 & 0 & 0 & 0 & 0 & -\sqrt{2}K_{22} \\ -K_{22} & K_{22} & 0 & 0 & 0 & 0 \\ 0 & 0 & 0 & 0 & 0 & 0 \end{pmatrix}$$

D_{6h}, D_6, C_{6v}, D_{3h}, D_{4h}, D_4, C_{4v}, D_{2d}

$$\begin{pmatrix} 0 & 0 & 0 & K_{14} & 0 & 0 \\ 0 & 0 & 0 & 0 & -K_{14} & 0 \\ 0 & 0 & 0 & 0 & 0 & 0 \end{pmatrix}$$

T, T_h, $O{:}T$, $T_d{:}T$, $O_h{:}T_h$

$$\begin{pmatrix} 0 & 0 & 0 & K_{14} & 0 & 0 \\ 0 & 0 & 0 & 0 & K_{14} & 0 \\ 0 & 0 & 0 & 0 & 0 & K_{14} \end{pmatrix}$$

These relations mean a considerable simplification of the entries of Table 8–3. In particular, as a consequence of the last relation in Eq. (15), $\mathbf{P} \equiv 0$ for the eight groups (D_{6h}, etc.) having the next to last form in Table 8–3 involving only the constant K_{14}. Others have a reduction in the number of independent constants.

Subsequent to this simplification, the two-index forms may be converted to three-index P_{ijk} by means of Eq. (7–2).

Problems

8–1. Explain why the existence of a pyromagnetic effect requires the presence of a spontaneous magnetization and vice versa. Also show that ferromagnetism or ferrimagnetism is possible only for the 31 crystal classes (Table 8–1) which admit pyromagnetism.

8–2. Obtain the form of a $T(2)^{ax}$ tensor for all the magnetic variants of groups D_{3d}.

8–3. Obtain the form of a $T(3)^{ax}$ tensor (the piezomagnetic tensor) for all the magnetic variants of the group D_{3h}.

8–4. How does the form of the $T(3)^{ax}$ tensor for magnetic group D_{3h}:D_3 change if the generating C_2-axis is taken to be the x-axis instead of the y-axis? (See Fig. 3–2.)

8–5. Prove the following statements:
(a) All axial c-tensors of even rank and polar c-tensors of odd rank are null for the 21 magnetic groups containing the inversion operation, i.
(b) All axial c-tensors of odd rank and polar c-tensors of even rank are null for the 21 classes containing the complementary operation $\underline{i}$.

8–6. Verify Eqs. (8–15).

9

Matter tensors of ranks 4 and 5

In dealing with matter tensors of rank 4, we see the considerable benefits of the present method, which considers group theoretically the 'force' and 'response' tensors, **X** and **Y** respectively, rather than looking directly for invariant components of the fourth-rank tensor **K** (see Section 4–6).

9–1 Relation between $T_S(2)$ and $T_S(2)$

In this section, we deal with a property K that couples two second-rank symmetric tensors of the type $T_S(2)$, the most important example of which is the elasticity tensor. Again we use the Weyl normalization to convert both **X** and **Y** to six-vectors, as in Eq. (4–4); **K** is then a 6×6 matrix and, in general, has 36 components. If, in addition, **K** is intrinsically symmetric (as in the case of elasticity) the matrix has a maximum of only 21 coefficients. If not (as, for example, for magnetoresistance) the full 36 coefficients must appear in the absence of crystal symmetry. These two cases can be handled together by treating the non-symmetric case and allowing the insertion of the symmetry requirement where appropriate.

After obtaining the group theoretical matrix $\bar{\mathbf{K}}$ and converting it back to $\mathbf{K} = (K_{ij})$, $i, j = 1, \ldots, 6$, we must then further convert the result either into four-index tensor quantities or into the two-index 'engineering' notation. To obtain the four-index quantities, K_{klmn}, we make use of Eq. (4–4) and obtain

$$K_{klmn} = \begin{Bmatrix} K_{ij} & \text{for } i, j = 1, 2, 3 \\ K_{ij}/\sqrt{2} & \text{for } i = 1, 2, 3 \text{ and } j = 4, 5, 6, \text{ or vice versa} \\ K_{ij}/2 & \text{for } i, j = 4, 5, 6 \end{Bmatrix} \tag{9–1}$$

This conversion is valid for any of the $T(4)$ tensors appropriate to this section. On the other hand, conversion of **K** to two-index engineering quantities depends on the exact nature of the **X** and **Y** tensors, and will be dealt with in subsequent sections.

In order to obtain $\bar{\mathbf{K}}$, we note that **X** and **Y** both generate the rep Γ_α. Non-vanishing coefficients will then be obtained by applying the fundamental theorem to each irrep into which Γ_α decomposes for the point group in question. Examination of the S-C-T tables (Appendix E) shows that the decomposition subdivides itself according to the Laue groups (see Section 6–4).

Let us begin with the *upper cubic crystals*, groups O_h, T_d and O. In each of these cases Γ_α takes the form

$$\Gamma_\alpha = A_1 + E + T \tag{9–2}$$

where $\alpha_1 + \alpha_2 + \alpha_3$ belongs to A_1, the pair $(2\alpha_1 - \alpha_2 - \alpha_3)$ and $(\alpha_2 - \alpha_3)$ belongs to E, and the triplet α_4, α_5 and α_6 to T. If we then take, for the symmetry coordinates $\bar{\mathbf{X}}$,

$$\left.\begin{array}{l} \bar{X}_1 = (X_1 + X_2 + X_3)/\sqrt{3} \\ \bar{X}_2 = (2X_1 - X_2 - X_3)/\sqrt{6}, \quad \bar{X}_3 = (X_2 - X_3)/\sqrt{2} \\ \bar{X}_4 = X_4,\ \bar{X}_5 = X_5,\ \bar{X}_6 = X_6 \end{array}\right\} \tag{9–3}$$

and the same for the $\bar{\mathbf{Y}}$-coordinates, we obtain for the matrix $\bar{\mathbf{K}}$, with the aid of the fundamental theorem,

$$\bar{\mathbf{K}} = \begin{pmatrix} A_1 & 0 & 0 & 0 & 0 & 0 \\ & E & 0 & 0 & 0 & 0 \\ & & E & 0 & 0 & 0 \\ & & & T & 0 & 0 \\ & & & & T & 0 \\ & & & & & T \end{pmatrix} \tag{9–4}$$

where only the subscripts are given for the non-zero components. Since this matrix is diagonal, it is the same whether or not **K** is intrinsically symmetric. Conversion to **K** gives the matrix in Table 9–1, in which

$$\left.\begin{array}{l} K_{11} = (K_A + 2K_E)/3 \\ K_{12} = (K_A - K_E)/3 \\ K_{44} = K_T \end{array}\right\} \tag{9–5}$$

At this point, it is desirable to digress to the even higher symmetry *isotropic system* which has the symmetry of the three-dimensional rotation group R(3). In the case of tensors of lower rank, a discussion of isotropic systems would not have been very useful. Note, especially, that for $T(2)$ tensors, the property for all cubic systems is already isotropic, so that a further increase in symmetry does not change the form of the tensor. Only for higher-order tensors, with larger numbers of coefficients, does a distinction between cubic and isotropic systems develop. Consideration of isotropic symmetry is of particular importance in the study of elasticity of glasses or of randomly oriented polycrystalline aggregates.

From the S-C-T table for R(3), we see that an isotropic system has the same symmetry coordinates as the upper cubic case, as given by Eq. (3), except that, now, $\bar{X}_2, \ldots, \bar{X}_6$ all belong to the five-dimensional irrep H. Thus, the form of $\bar{\mathbf{K}}$ is the same as Eq. (4), except that K_E and K_T are now equal and are replaced by K_H. Accordingly, we obtain the same **K** matrix, with the additional condition that

$$K_{44} = K_{11} - K_{12} \quad \text{(isotropic)} \tag{9–6}$$

that is, there are now only two independent K_{ij} constants.

The *lower cubic systems*, T_h and T, differ from the upper cubics in that

$$\Gamma_\alpha = \mathrm{A} + \tilde{\mathrm{E}} + \mathrm{T_g} \tag{9–7}$$

where the complex symmetry coordinates of $\tilde{\mathrm{E}}$ are $(2\alpha_1 - \alpha_2 - \alpha_3) \mp \mathrm{i}(\alpha_2 - \alpha_3)$. Therefore, using the same symmetry coordinates as in Eq. (3), the matrix of Eq. (4) is obtained, except that the 2×2 diagonal part involving K_E is replaced by the block

$$\begin{pmatrix} \tilde{\mathrm{E}}^{\mathrm{re}} & \tilde{\mathrm{E}}^{\mathrm{im}} \\ -\tilde{\mathrm{E}}^{\mathrm{im}} & \tilde{\mathrm{E}}^{\mathrm{re}} \end{pmatrix}$$

in accordance with Eqs. (4–22, 4–23). Upon transformation to the original coordinates we readily obtain the matrix **K** as in Table 9–1, where now

$$\left.\begin{aligned} K_{11} &= (K_A + 2K_{\tilde{E}}^{\mathrm{re}})/3 \\ K_{12} &= (K_A - K_{\tilde{E}}^{\mathrm{re}} + \sqrt{3}K_{\tilde{E}}^{\mathrm{im}})/3 \\ K_{13} &= (K_A - K_{\tilde{E}}^{\mathrm{re}} - \sqrt{3}K_{\tilde{E}}^{\mathrm{im}})/3 \\ K_{44} &= K_T \end{aligned}\right\} \tag{9–8}$$

This set involves one more constant than the upper cubic case. However, if **K** is intrinsically symmetric, $K_{\tilde{E}}^{\mathrm{im}} = 0$, and the **K** matrix reverts to that of the previous case; hence, for symmetric **K** all cubic systems are alike.

Table 9–2 gives the symmetric versions of **K**, that is, for a $T_S(4)$ tensor, for the higher symmetry crystals. Actually, these results are very easily obtained from Table 9–1 by inspection, but the case of $T_S(4)$ is so important that we feel it very desirable to provide a separate table.

We next turn to *uniaxial crystals*. For all these cases both $\alpha_1 + \alpha_2$ and α_3 belong to an A irrep. Accordingly, we take for the symmetry coordinates $\bar{X}_1, \ldots, \bar{X}_6$ the set

$$(X_1 + X_2)/\sqrt{2},\ (X_1 - X_2)/\sqrt{2},\ X_3,\ X_4,\ X_5,\ X_6 \tag{9–9}$$

and similarly for the $\bar{\mathbf{Y}}$ coordinates.

It is convenient to begin the uniaxials with the *upper tetragonal classes* (D_{4h}, D_{2d}, D_{4v} and D_4) for which

$$\Gamma_\alpha = 2A + B_1 + B_2 + E \tag{9–10}$$

From the S-C-T table we see that $\alpha_1 - \alpha_2$ belongs to B_1, α_6 to B_2, and the pair (α_4, α_5) to E. Thus, we obtain $\bar{\mathbf{K}}$ in the form

$$\bar{\mathbf{K}} = \begin{pmatrix} A,11 & 0 & A,12 & 0 & 0 & 0 \\ 0 & B_1 & 0 & 0 & 0 & 0 \\ A,21 & 0 & A,22 & 0 & 0 & 0 \\ 0 & 0 & 0 & E & 0 & 0 \\ 0 & 0 & 0 & 0 & E & 0 \\ 0 & 0 & 0 & 0 & 0 & B_2 \end{pmatrix} \tag{9–11}$$

with seven independent constants, reduced to six if **K** is symmetric. Transforming to $\mathbf{K} = (K_{ij})$ we obtain the matrix given in Table 9–1, with

$$\left.\begin{aligned} &K_{11} = (K_{A,11} + K_{B_1})/2 \\ &K_{12} = (K_{A,11} - K_{B_1})/2 \\ &K_{13} = K_{A,12}/\sqrt{2}; \quad K_{31} = K_{A,21}/\sqrt{2}; \quad K_{33} = K_{A,22} \\ &K_{44} = K_E; \quad K_{66} = K_{B_2} \end{aligned}\right\} \tag{9–12}$$

For the symmetric case, $K_{13} = K_{31}$, as in Table 9–2.

We turn next to the *upper hexagonal classes* (D_{6h}, D_{3h}, C_{6v} and D_6), for which

$$\Gamma_\alpha = 2A + E_1 + E_2 \tag{9–13}$$

(for D_{3h}, E_1 and E_2 become E″ and E′, respectively). The $\bar{\mathbf{K}}$ matrix takes the same form as Eq. (11) except that E now becomes E_1, while B_1 and B_2 are replaced by E_2. Thus there is one less constant than for the tetragonal case, that is, six independent constants if **K** is not symmetric and five constants if it is symmetric. The conversion to K_{ij} is then similar to Eq. (12) with the replacements of B_1 and B_2 by E_2 and E by E_1. The **K** matrix given in Tables 9–1 and 9–2 is then the same as for the upper tetragonal, except that

$$K_{66} = K_{11} - K_{12} \tag{9–14}$$

Equation (14) represents the reduction in the number of constants by one in going from the tetragonal to this hexagonal case.

The *upper trigonal classes* (D_{3d}, D_3 and C_{3v}) are similar to the hexagonal but have

$$\Gamma_\alpha = 2A + 2E \tag{9–15}$$

that is, instead of two different E irreps, there is only one (repeated). Accordingly, we must concern ourselves with similarity of orientation.

Inspection of the S-C-T tables shows that the orientations of the two pairs of coordinates can always be taken as: $(\alpha_1 - \alpha_2, \alpha_6)$ and (α_4, α_5). Therefore, defining our symmetry coordinates again as in Eq. (9) we obtain the following $\bar{\mathbf{K}}$ matrix:

$$\bar{\mathbf{K}} = \begin{pmatrix} \mathrm{A,11} & 0 & \mathrm{A,12} & 0 & 0 & 0 \\ 0 & \mathrm{E,22} & 0 & \mathrm{E,21} & 0 & 0 \\ \mathrm{A,21} & 0 & \mathrm{A,22} & 0 & 0 & 0 \\ 0 & \mathrm{E,12} & 0 & \mathrm{E,11} & 0 & 0 \\ 0 & 0 & 0 & 0 & \mathrm{E,11} & \mathrm{E,12} \\ 0 & 0 & 0 & 0 & \mathrm{E,21} & \mathrm{E,22} \end{pmatrix} \quad (9\text{–}16)$$

which involves eight independent constants (six if symmetric). The **K** matrix is given in Tables 9–1 and 9–2. Here the K_{ij} coefficients are the same as for the upper hexagonal case, replacing E_1 by E,11 and E_2 by E,22, with the following additional coefficients:

$$K_{14} = K_{\mathrm{E,21}}/\sqrt{2}; \quad K_{41} = K_{\mathrm{E,12}}/\sqrt{2} \quad (9\text{–}17)$$

It is worth noting here that the reason that all upper classes for a given major symmetry axis (tetragonal, hexagonal or trigonal) fall together so neatly for the present $T(4)$ tensor is a consequence of our earlier choice of the pair (α_4, α_5) to set the orientation of the E irreps in these cases (see Section 3–5). If we had set the orientation with (x, y) instead of (α_4, α_5), the same results would have been obtained, but in a more untidy manner.

We turn now to the lower symmetry uniaxial crystals, beginning with the *lower tetragonal classes* (C_{4h},C_4 and S_4) for which

$$\Gamma_\alpha = 2\mathrm{A} + 2\mathrm{B} + \tilde{\mathrm{E}} \quad (9\text{–}18)$$

with $\alpha_1 - \alpha_2$ and α_6 both belonging to B and $\alpha_4 \pm i\alpha_5$ to $\tilde{\mathrm{E}}$. This case differs from the upper tetragonal by a repeated B irrep replacing B_1 and B_2 and by a complex $\tilde{\mathrm{E}}$ irrep replacing E. Using the same symmetry coordinates, Eq. (9), we readily obtain, for the $\bar{\mathbf{K}}$ matrix,

$$\bar{\mathbf{K}} = \begin{pmatrix} \mathrm{A,11} & 0 & \mathrm{A,12} & 0 & 0 & 0 \\ 0 & \mathrm{B,11} & 0 & 0 & 0 & \mathrm{B,12} \\ \mathrm{A,21} & 0 & \mathrm{A,22} & 0 & 0 & 0 \\ 0 & 0 & 0 & \tilde{\mathrm{E}}^{\mathrm{re}} & -\tilde{\mathrm{E}}^{\mathrm{im}} & 0 \\ 0 & 0 & 0 & \tilde{\mathrm{E}}^{\mathrm{im}} & \tilde{\mathrm{E}}^{\mathrm{re}} & 0 \\ 0 & \mathrm{B,21} & 0 & 0 & 0 & \mathrm{B,22} \end{pmatrix} \quad (9\text{–}19)$$

which involves ten independent constants if **K** is non-symmetric and only seven if it is symmetric (since, then, A,21 = A,12 and B,21 = B,12, while $\tilde{\mathrm{E}}^{\mathrm{im}} = 0$). The respective **K** matrix is given in Tables 9–1 and 9–2 with constants K_{ij} as given in Eq. (12), except that B,11 and B,22 replace B_1

and B_2, respectively, and $\tilde{E}^{re}$ replaces E. The additional three constants are the following:

$$\left.\begin{aligned} K_{16} &= -K_{26} = K_{B,12}/\sqrt{2} \\ K_{61} &= -K_{62} = K_{B,21}/\sqrt{2} \\ K_{45} &= -K_{54} = K_{\tilde{E}}^{im} \end{aligned}\right\} \qquad (9\text{–}20)$$

For the *lower hexagonal classes* (C_{6h}, C_{3h}, C_6) we have

$$\Gamma_{\alpha} = 2A_1 + \tilde{E}_1 + \tilde{E}_2 \qquad (9\text{–}21)$$

(except for C_{3h} which has $2A_1 + \tilde{E}'' + \tilde{E}'$ instead), with $\alpha_4 \pm i\alpha_5$ belonging to $\tilde{E}_1$ and $\alpha_1 - \alpha_2 \mp i\sqrt{2}\alpha_6$ to $\tilde{E}_2$. Here $\bar{\mathbf{K}}$ becomes

$$\bar{\mathbf{K}} = \begin{pmatrix} A,11 & 0 & A,12 & 0 & 0 & 0 \\ 0 & \tilde{E}_2^{re} & 0 & 0 & 0 & \tilde{E}_2^{im} \\ A,21 & 0 & A,22 & 0 & 0 & 0 \\ 0 & 0 & 0 & \tilde{E}_1^{re} & -\tilde{E}_1^{im} & 0 \\ 0 & 0 & 0 & \tilde{E}_1^{im} & \tilde{E}_1^{re} & 0 \\ 0 & -\tilde{E}_2^{im} & 0 & 0 & 0 & \tilde{E}_2^{re} \end{pmatrix} \qquad (9\text{–}22)$$

Thus, there are eight independent constants, but only five if **K** is symmetric. The results for **K** are given in Tables 9–1 and 9–2, with K_{11}, K_{12}, K_{13}, K_{31}, K_{33}, K_{44} and K_{66} as for the upper hexagonals (except that E_1 is replaced by $\tilde{E}_1^{re}$ and E_2 by $\tilde{E}_2^{re}$. The two additional constants are

$$\left.\begin{aligned} K_{16} &= -K_{26} = K_{62} = -K_{61} = K_{\tilde{E}_2}^{im}/\sqrt{2} \\ K_{45} &= -K_{54} = -K_{\tilde{E}_1}^{im} \end{aligned}\right\} \qquad (9\text{–}23)$$

The last of our uniaxial classes are the *lower trigonal* (S_6, C_3) for which

$$\Gamma_{\alpha} = 2A + 2\tilde{E} \qquad (9\text{–}24)$$

with $\alpha_4 \pm i\alpha_5$ and $(\alpha_1 - \alpha_2) \pm i\sqrt{2}\alpha_6$ belonging to $\tilde{E}$. This case of a repeated $\tilde{E}$ irrep is dealt with in Section 4–5 and leads to the following $\bar{\mathbf{K}}$ matrix:

$$\bar{\mathbf{K}} = \begin{pmatrix} A,11 & 0 & A,12 & 0 & 0 & 0 \\ 0 & \tilde{E},22^{re} & 0 & \tilde{E},21^{re} & -\tilde{E},21^{im} & -\tilde{E},22^{im} \\ A,21 & 0 & A,22 & 0 & 0 & 0 \\ 0 & \tilde{E},12^{re} & 0 & \tilde{E},11^{re} & -\tilde{E}11^{im} & -\tilde{E},12^{im} \\ 0 & \tilde{E},12^{im} & 0 & \tilde{E},11^{im} & \tilde{E},11^{re} & \tilde{E},12^{re} \\ 0 & \tilde{E},22^{im} & 0 & \tilde{E},21^{im} & \tilde{E},21^{re} & \tilde{E},22^{re} \end{pmatrix} \qquad (9\text{–}25)$$

involving 12 constants, but only seven if symmetric. The **K** matrix is given in Table 9–1, in which the K_{ij} are the same as for the upper trigonal case with the following additions:

$$\left.\begin{aligned} K_{61} &= -K_{62} = -K_{16} = K_{26} = K^{\text{im}}_{\tilde{\text{E}},22}/\sqrt{2} \\ K_{45} &= -K_{54} = -K^{\text{im}}_{\tilde{\text{E}},11} \\ K_{51} &= -K_{52} = -K_{46}/\sqrt{2} = K^{\text{im}}_{\tilde{\text{E}},12}/\sqrt{2} \\ K_{15} &= -K_{25} = -K_{64}/\sqrt{2} = -K^{\text{im}}_{\tilde{\text{E}},21}/\sqrt{2} \end{aligned}\right\} \tag{9–26}$$

The *orthorhombic classes*, D_{2h} and D_2, have

$$\Gamma_\alpha = 3A + B_1 + B_2 + B_3 \tag{9–27}$$

while C_{2v} has $3A_1 + A_2 + B_1 + B_2$. Here α_1, α_2 and α_3 all belong to the triply repeated A irrep, while α_6, α_5 and α_4 in turn belong to the three one-dimensional non-repeating irreps. We therefore use the original coordinates $\alpha_1, \ldots, \alpha_6$ as the symmetry coordinates. It is then straightforward to obtain the **K** matrix in Table 9–1.

Finally, for *monoclinic classes* C_{2h}, C_2 and C_s, Γ_α is always of the form

$$\Gamma_\alpha = 4A + 2B \tag{9–28}$$

where α_1, α_2, α_3 and α_6 belong to A and α_4 and α_5 to B. Again, the results of Table 9–1 are obtained directly.

9–2 Application to the elastic constants

In Table 9–2, we have already given the K_{ij} tables for the intrinsically symmetric case of a $T(4)$ tensor. The most important example of this tensor type is that of the elastic constants, **s** or **c** (see Section 1–3). It is important to convert the results of this table to the more practical 'engineering' constants.

Conversion to the four-index notation has already been given in Eq. (1). When that equation is combined with the definition of the engineering stresses and strains (see Section 1–4) we can arrive at the direct conversion of the symmetric K_{ij} of Table 9–2 into either the $\mathbf{c}^e$ or $\mathbf{s}^e$ matrices. This is readily shown to be as follows. For the elastic compliance matrix **s**:

$$s^{\text{e}}_{ij} = f s_{ij} \tag{9–29}$$

where s_{ij} has the form given by Table 9–2, and

$$\begin{aligned} f &= 1 && \text{if } i, j \text{ are } 1, 2 \text{ or } 3 \\ &= 2 && \text{if } i, j \text{ are } 4, 5 \text{ or } 6 \\ &= \sqrt{2} && \text{if } i = 1, 2 \text{ or } 3,\ j = 4, 5 \text{ or } 6 \end{aligned}$$

Table 9–1. *Forms of a* T(4) *tensor that couples two* $T_S(2)$ *tensors, for various crystal classes*

Triclinic (36 constants)

K_{11}	K_{12}	K_{13}	K_{14}	K_{15}	K_{16}
K_{21}	K_{22}	K_{23}	K_{24}	K_{25}	K_{26}
K_{31}	K_{32}	K_{33}	K_{34}	K_{35}	K_{36}
K_{41}	K_{42}	K_{43}	K_{44}	K_{45}	K_{46}
K_{51}	K_{52}	K_{53}	K_{54}	K_{55}	K_{56}
K_{61}	K_{62}	K_{63}	K_{64}	K_{65}	K_{66}

Monoclinic (20)

K_{11}	K_{12}	K_{13}	0	0	K_{16}
K_{21}	K_{22}	K_{23}	0	0	K_{26}
K_{31}	K_{32}	K_{33}	0	0	K_{36}
0	0	0	K_{44}	K_{45}	0
0	0	0	K_{54}	K_{55}	0
K_{61}	K_{62}	K_{63}	0	0	K_{66}

Orthorhombic (12)

K_{11}	K_{12}	K_{13}	0	0	0
K_{21}	K_{22}	K_{23}	0	0	0
K_{31}	K_{32}	K_{33}	0	0	0
0	0	0	K_{44}	0	0
0	0	0	0	K_{55}	0
0	0	0	0	0	K_{66}

Lower tetragonal (10)

K_{11}	K_{12}	K_{13}	0	0	K_{16}
K_{12}	K_{11}	K_{13}	0	0	$-K_{16}$
K_{31}	K_{31}	K_{33}	0	0	0
0	0	0	K_{44}	K_{45}	0
0	0	0	$-K_{45}$	K_{44}	0
K_{61}	$-K_{61}$	0	0	0	K_{66}

Upper tetragonal (7)

K_{11}	K_{12}	K_{13}	0	0	0
K_{12}	K_{11}	K_{13}	0	0	0
K_{31}	K_{31}	K_{33}	0	0	0
0	0	0	K_{44}	0	0
0	0	0	0	K_{44}	0
0	0	0	0	0	K_{66}

Lower trigonal (12)

K_{11}	K_{12}	K_{13}	K_{14}	K_{15}	$-K_{61}$
K_{12}	K_{11}	K_{13}	$-K_{14}$	$-K_{15}$	K_{61}
K_{31}	K_{31}	K_{33}	0	0	0
K_{41}	$-K_{41}$	0	K_{44}	K_{45}	$-\sqrt{2}K_{51}$
K_{51}	$-K_{51}$	0	$-K_{45}$	K_{44}	$\sqrt{2}K_{41}$
K_{61}	$-K_{61}$	0	$-\sqrt{2}K_{15}$	$\sqrt{2}K_{14}$	$(K_{11} - K_{12})$

Upper trigonal (8)

$$\begin{matrix} K_{11} & K_{12} & K_{13} & K_{14} & 0 & 0 \\ K_{12} & K_{11} & K_{13} & -K_{14} & 0 & 0 \\ K_{31} & K_{31} & K_{33} & 0 & 0 & 0 \\ K_{41} & -K_{41} & 0 & K_{44} & 0 & 0 \\ 0 & 0 & 0 & 0 & K_{44} & \sqrt{2}K_{41} \\ 0 & 0 & 0 & 0 & \sqrt{2}K_{14} & (K_{11} - K_{12}) \end{matrix}$$

Lower hexagonal (8)

$$\begin{matrix} K_{11} & K_{12} & K_{13} & 0 & 0 & -K_{61} \\ K_{12} & K_{11} & K_{13} & 0 & 0 & K_{61} \\ K_{31} & K_{31} & K_{33} & 0 & 0 & 0 \\ 0 & 0 & 0 & K_{44} & K_{45} & 0 \\ 0 & 0 & 0 & -K_{45} & K_{44} & 0 \\ K_{61} & -K_{61} & 0 & 0 & 0 & (K_{11} - K_{12}) \end{matrix}$$

Upper hexagonal (6)

$$\begin{matrix} K_{11} & K_{12} & K_{13} & 0 & 0 & 0 \\ K_{12} & K_{11} & K_{13} & 0 & 0 & 0 \\ K_{31} & K_{31} & K_{33} & 0 & 0 & 0 \\ 0 & 0 & 0 & K_{44} & 0 & 0 \\ 0 & 0 & 0 & 0 & K_{44} & 0 \\ 0 & 0 & 0 & 0 & 0 & (K_{11} - K_{12}) \end{matrix}$$

Lower cubic (4)

$$\begin{matrix} K_{11} & K_{12} & K_{13} & 0 & 0 & 0 \\ K_{13} & K_{11} & K_{12} & 0 & 0 & 0 \\ K_{12} & K_{13} & K_{11} & 0 & 0 & 0 \\ 0 & 0 & 0 & K_{44} & 0 & 0 \\ 0 & 0 & 0 & 0 & K_{44} & 0 \\ 0 & 0 & 0 & 0 & 0 & K_{44} \end{matrix}$$

Upper cubic (3)

$$\begin{matrix} K_{11} & K_{12} & K_{12} & 0 & 0 & 0 \\ K_{12} & K_{11} & K_{12} & 0 & 0 & 0 \\ K_{12} & K_{12} & K_{11} & 0 & 0 & 0 \\ 0 & 0 & 0 & K_{44} & 0 & 0 \\ 0 & 0 & 0 & 0 & K_{44} & 0 \\ 0 & 0 & 0 & 0 & 0 & K_{44} \end{matrix}$$

Isotropic (2)

Same as upper cubic but with

$$K_{44} = K_{11} - K_{12}$$

Table 9–2. *Forms of a* $T_S(4)$ *tensor for some of the higher crystal symmetries (others can be obtained from Table 9–1 by setting* $K_{ij} = K_{ji}$*) (To use for elastic compliances and stiffness refer to Eqs. (29) and (30) respectively)*

All cubics

K_{11}	K_{12}	K_{12}	0	0	0
	K_{11}	K_{12}	0	0	0
		K_{11}	0	0	0
			K_{44}	0	0
				K_{44}	0
					K_{44}

(For isotropic case, set $K_{44} = K_{11} - K_{12}$)

All hexagonals

K_{11}	K_{12}	K_{13}	0	0	0
	K_{11}	K_{13}	0	0	0
		K_{33}	0	0	0
			K_{44}	0	0
				K_{44}	0
					$(K_{11} - K_{12})$

Upper tetragonal

K_{11}	K_{12}	K_{13}	0	0	0
	K_{11}	K_{13}	0	0	0
		K_{33}	0	0	0
			K_{44}	0	0
				K_{44}	0
					K_{66}

Lower tetragonal

K_{11}	K_{12}	K_{13}	0	0	K_{16}
	K_{11}	K_{13}	0	0	$-K_{16}$
		K_{33}	0	0	0
			K_{44}	0	0
				K_{44}	0
					K_{66}

Upper trigonal

K_{11}	K_{12}	K_{13}	K_{14}	0	0
	K_{11}	K_{13}	$-K_{14}$	0	0
		K_{33}	0	0	0
			K_{44}	0	0
				K_{44}	$\sqrt{2}K_{14}$
					$(K_{11} - K_{12})$

For the **c** matrix:

$$c^{e}_{ij} = gc_{ij} \qquad (9\text{–}30)$$

where

$$\begin{aligned} g &= 1 && \text{if } i, j \text{ are } 1, 2 \text{ or } 3 \\ &= 1/2 && \text{if } i, j \text{ are } 4, 5 \text{ or } 6 \\ &= 1/\sqrt{2} && \text{if } i = 1, 2 \text{ or } 3,\ j = 4, 5 \text{ or } 6 \end{aligned}$$

The same results can conveniently be shown by giving the multiplying factors for the four 3 × 3 blocks of coefficients that make up the matrices of Table 9–2, as follows:

$$\mathbf{s}^{e} = \left(\begin{array}{c|c} \times 1 & \times\sqrt{2} \\ \hline \times\sqrt{2} & \times 2 \end{array}\right); \quad \mathbf{c}^{e} = \left(\begin{array}{c|c} \times 1 & \times 1/\sqrt{2} \\ \hline \times 1/\sqrt{2} & \times 1/2 \end{array}\right) \qquad (9\text{–}31)$$

Elastic constants are almost never determined by static methods, that is, by the application of a constant stress and observation of the appropriate strains. Such methods would have very limited precision, especially when plastic deformation sets in at higher stress levels. Instead, dynamical methods are generally used. There are two basic dynamical methods for obtaining the elastic constants. The older method uses resonant bars and obtains the coefficients s_{ij} from appropriate resonant frequencies. In all such methods, a resonant frequency, f, is related to an effective constant, s_{eff}, by

$$s_{\text{eff}} \propto \rho f^2 \qquad (9\text{–}32)$$

where ρ is the density, and the proportionality constant depends on the dimensions of the sample and on the type of vibration it undergoes. (See e.g. Nowick and Berry, 1972.) The most important types are longitudinal vibration of a crystal bar along the z'-axis, where $s_{\text{eff}} = s'_{33}$, and torsional vibration, for which $s_{\text{eff}} = 1/2(s'_{44} + s'_{55})$. Here the s'_{ij} are the elastic compliance constants in the coordinate system of the axes of the bar. Knowing the four-index tensor forms of the elastic constants in standard coordinates makes possible the calculation of the constants in a crystal with rotated axes. For cubic crystals the results are (in terms of the two-index engineering moduli)

$$s'_{33} = s^{e}_{11} - 2[s^{e}_{11} - s^{e}_{12} - (1/2)s^{e}_{44}]\Gamma \qquad (9\text{–}33)$$

where

$$\Gamma = \gamma_1^2\gamma_2^2 + \gamma_2^2\gamma_3^2 + \gamma_3^2\gamma_1^2 \qquad (9\text{–}34)$$

and γ_1, γ_2 and γ_3 are the direction cosines between the sample axis (z')

and the three crystal axes. The quantity Γ varies from zero for z' parallel to a $\langle 100\rangle$ direction, to a maximum of 1/3 for z' in a $\langle 111\rangle$ direction. Similarly, for torsional vibration about the z'-axis, the result is

$$s_{\text{eff}} = s^{\text{e}}_{44} + 4[s^{\text{e}}_{11} - s^{\text{e}}_{12} - (1/2)s^{\text{e}}_{44}]\Gamma \qquad (9\text{–}35)$$

Note that for an isotropic material, the quantity in brackets in Eqs. (33) and (35) vanishes, and the appropriate modulus becomes independent of orientation. Similar relations exist for other crystal symmetries (Nowick and Berry, 1972). The above relations apply only under free (unconstrained) vibration. For arbitrary orientations, even of a cubic crystal, all coefficients s_{ij}, including those for which $i \leqslant 3$, $j > 3$, may occur. The latter coefficients relate an extensional strain to a shear stress, or vice versa, meaning that extension is accompanied by shear or torsion by flexure. Accordingly, if the samples are constrained to vibrate in a 'pure' mode, corrections to the effective moduli must be made (Hearmon, 1946). In any case, it is best to choose high-symmetry orientations of the sample where these couplings vanish (although such a choice may not always be possible).

Modern studies of elastic constants have used primarily ultrasonic wave propagation methods (Truell *et al.*, 1969). Such methods involve the elastic stiffness constants, c_{ij}. Most commonly, a *pulse propagation method* is used. This method employs a pulse or wave packet (of about ten wavelengths) whose length is small compared to the specimen length in the direction of propagation. A series of such pulses is generated by an appropriate transducer bonded to one end of the sample. The velocity of the pulses can be determined from the time for them to reach the other end. Alternatively, it may be obtained from the time taken to return to the starting point after suffering a reflection at the free end; in this way, a single transducer can be used, as illustrated schematically in Fig. 9–1. The type of propagation which lends itself most easily to analysis is that of a

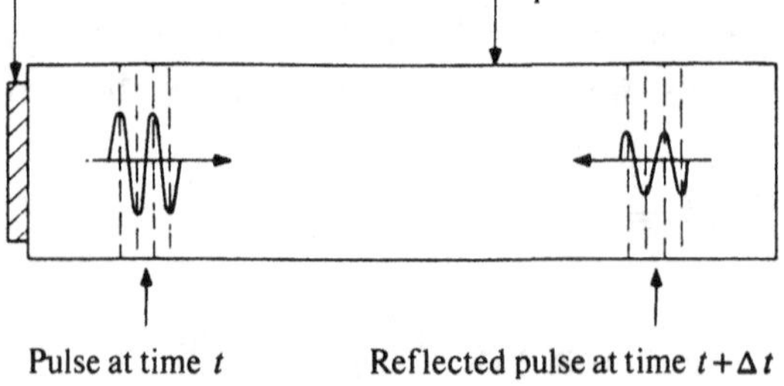

Fig. 9–1. Schematic illustration of the ultrasonic pulse method.

plane (non-spreading) harmonic wave traveling along the z'-direction in an elastic medium. The particle displacement, u, may be in the z'-direction (longitudinal wave) or normal to it (transverse wave). Such cases may be expressed by the one-dimensional wave equation of the form

$$\rho\frac{\partial^2 u}{\partial t^2} = c_{\text{eff}}\frac{\partial^2 u}{\partial z'^2} \tag{9–36}$$

with an effective elastic constant c_{eff}. A complicating factor is the finite length of the wave.

In an isotropic medium, for a given direction of propagation, three distinct waves can be generated. The first is longitudinal, with velocity v_l and the other two are transverse, both with velocity v_{t}. The velocities are related to c_{eff} by

$$\rho v^2 = c_{\text{eff}} \tag{9–37}$$

In a crystal, there are still three waves for a given propagation direction, but for an arbitrary choice of z' these waves will involve a mixture of modes of vibration. Only for high-symmetry orientations of the crystal are separate longitudinal and transverse modes maintained; such orientations should be used whenever possible.

The simplest cases are those involving only a single group-theoretical coefficient c_γ or two coefficients of which one belongs to the A_1 irrep. We give such expressions for v_l and v_{t} (Nowick and Berry, 1972) in which subscripts give the crystal direction of propagation and, for transverse waves, a superscript shows the polarization direction (if this needs to be specified). For *cubic crystals*:

$$(\rho v_l^2)_{\langle 100\rangle} = \frac{1}{3}(c_{\text{A}} + 2c_{\text{E}}) = c_{11}^{\text{e}} \tag{9–38}$$

$$(\rho v_{\text{t}}^2)_{\langle 100\rangle} = (\rho v_{\text{t}}^2)_{[110]}^{[001]} = \frac{1}{2}c_{\text{T}} = c_{44}^{\text{e}} \tag{9–39}$$

$$(\rho v_{\text{t}}^2)_{[110]}^{[1\bar{1}0]} = \frac{1}{2}c_{\text{E}} = \frac{1}{2}(c_{11}^{\text{e}} - c_{12}^{\text{e}}) \tag{9–40}$$

$$(\rho v_l^2)_{\langle 111\rangle} = \frac{1}{3}(c_{\text{A}} + 2c_{\text{T}}) = \frac{1}{3}(c_{11}^{\text{e}} + 2c_{12}^{\text{e}} + 4c_{44}^{\text{e}}) \tag{9–41}$$

For hexagonal crystals, the following relations are obtained:

$$(\rho v_l^2)_{[001]} = c_{\text{A},22} = c_{33}^{\text{e}} \tag{9–42}$$

$$(\rho v_{\text{t}}^2)_{[001]} = \frac{1}{2}c_{\text{E}_1} = c_{44}^{\text{e}} \tag{9–43}$$

$$(\rho v_l^2)_{[hk0]} = \frac{1}{2}(c_{\text{A},11} + c_{\text{E}_2}) = c_{11}^{\text{e}} \tag{9–44}$$

Table 9–3. *Elastic compliance constants* c_{ij} *(in* 10^{11} *dyn/cm²) and anisotropy factor,* A, *of some cubic crystals near room temperature (from Landolt–Börnstein tables)*

Crystal	$\overset{e}{c}_{11}$	$\overset{e}{c}_{12}$	$\overset{e}{c}_{44}$	A
Ag	12.4	9.34	4.61	3.0
Na	0.73	0.63	0.42	7.8
Si	16.6	6.39	7.96	1.56
NaCl	4.87	1.24	1.26	0.7
MgO	28.9	8.8	15.5	1.54
CaF_2	16.5	4.7	3.39	0.57

$$(\rho v_t^2)^{[k\bar{h}0]}_{[hk0]} = \frac{1}{2} c_{E_2} = \frac{1}{2}(\overset{e}{c}_{11} - \overset{e}{c}_{12}) \tag{9–45}$$

$$(\rho v_t^2)^{[001]}_{[hk0]} = \frac{1}{2} c_{E_1} = \overset{e}{c}_{44} \tag{9–46}$$

To determine the remaining constant, $\overset{e}{c}_{13}$, is more difficult, requiring propagation in a direction of lower symmetry.

In a similar way, expressions for wave velocity in terms of c_{ij} elastic constants may be obtained for other crystal structures (Truell *et al.*, 1969).

Values of the principal elastic stiffness constants for a number of cubic crystals are given in Table 9–3. Also included is the anisotropy factor

$$A = \frac{c_T}{c_E} = 2\frac{\overset{e}{c}_{44}}{(\overset{e}{c}_{11} - \overset{e}{c}_{12})} \tag{9–47}$$

This is the factor that becomes equal to unity for an isotropic material. (See Eq. (6).)

The reader should be aware that, throughout the literature on elasticity, two-index constants c_{ij} or s_{ij} refer to what we have called the engineering constants $\overset{e}{c}_{ij}$ and $\overset{e}{s}_{ij}$. For the purposes of this book, however, we have reserved the two-index constants without the superscript e to refer to relations between stress and strain expressed as six-vectors.

Finally, we discuss the interconversion between stiffness coefficients, c_{ij}, and compliances, s_{ij}. From Eqs. (1–12) and (1–13), we see that the **s** and **c** matrices are reciprocals of each other. Considering that they are 6×6 matrices, carrying out the interconversion is indeed a messy task for low-symmetry crystals; however, it becomes much easier as the symmetry increases. The process is greatly simplified by making use of any diagonal elements in the symmetrized $\bar{\mathbf{K}}$ tensor. In the cubic case, where the $\bar{\mathbf{K}}$ tensor is completely diagonal, Eq. (4), the conversion is immediate, as follows:

$$s_A = \frac{1}{c_A}; \quad s_E = \frac{1}{c_E}; \quad s_T = \frac{1}{c_T} \tag{9–48}$$

so that

$$s^e_{11} + 2s^e_{12} = (c^e_{11} + 2c^e_{12})^{-1} \tag{9–49}$$

$$s^e_{11} - s^e_{12} = (c^e_{11} - c^e_{12})^{-1} \tag{9–50}$$

$$s^e_{44} = (c^e_{44})^{-1} \tag{9–51}$$

It is also important to note that the three group-theoretical constants in cubic crystals have special significance (see Problem 9–7). The constant $c_A/3$ is readily shown to equal the bulk modulus, while c_E and c_T are two independent shear constants (which become equal to each other in the isotropic case).

Returning to the question of interconversion between the c^e_{ij} and s^e_{ij} constants, for crystals of lower-than-cubic symmetry, the presence of diagonal elements in the $\bar{\mathbf{K}}$ tensor greatly aids the process, leaving smaller determinantal equations to complete the inversion.

9–3 Some applications of non-symmetric $T(4)$ tensors

While the elastic constants involve the symmetric tensors given by Table 9–2, there are several important properties which are described by the non-symmetric tensors of Table 9–1. Again, we note that, whereas the conversion from the two-index (six-vector) notation used in this chapter to four-index (tensor) notation is universal, as given by Eq. (1), the conversion to two-index engineering notation is different for each specific property in question.

For the photoelastic tensor $\mathbf{q}$, Eq. (1–31), $\mathbf{X}$ is the stress and $\mathbf{Y}$ represents the change in the β_{ij} coefficients of the index ellipsoid. The relation between the two-index engineering values and the four-index tensor components is given by Eq. (1–35). Combining with Eq. (1) we obtain the conversion of the matrices of Table 9–1 to engineering notation. Expressed in the same block form as Eq. (31) we have

$$\mathbf{q}^e = \left(\begin{array}{c|c} \times 1 & \times\sqrt{2} \\ \hline \times 1/\sqrt{2} & \times 1 \end{array}\right) \tag{9–52}$$

For the strain-optical tensor **m**, Eq. (1–32), on the other hand, we find

$$\mathbf{m}^{e} = \left(\begin{array}{c|c} \times 1 & \times 1/\sqrt{2} \\ \hline \times 1/\sqrt{2} & \times 1/2 \end{array}\right) \tag{9–53}$$

while for the electrostrictive tensor γ, Eq. (1–27):

$$\gamma^{e} = \left(\begin{array}{c|c} \times 1 & \times \sqrt{2} \\ \hline \times \sqrt{2} & \times 2 \end{array}\right) \tag{9–54}$$

The magnetoresistance ξ, Eq. (2–26), as well as the magnetothermal conductivity and piezoresistance, Eq. (2–16), have identical schemes as the photoelastic tensor **q**, namely, Eq. (52).

The photoelastic tensor may be regarded as involving a lowering of the symmetry of the impermeability tensor $\boldsymbol{\beta}$ by the application of a stress σ, since

$$\beta_{ij} = \beta^{0}_{ij} + q_{ijkl}\sigma_{kl} \tag{9–55}$$

For example, for an upper cubic crystal, for which $\boldsymbol{\beta}^{0}$ is a constant tensor, the application of a stress along the [100] direction converts the crystal from an optically isotropic to a uniaxial (birefringent) crystal, with optic axis along [100]. If we revert to single-index engineering notation for β and σ and note that, since $\beta = 1/n^2$, a change in β corresponds to a change in refractive index

$$\delta\beta = \delta(1/n^2) = -(2/n_0^3)\delta n \tag{9–56}$$

where n_0 is the refractive index in the absence of stress. Thus, we have in the [100] direction

$$n_1 = n_0 - q^{e}_{11} n_0^3 \sigma/2 \tag{9–57}$$

and in the other two cube directions

$$n_2 = n_3 = n_0 - q^{e}_{12} n_0^3 \sigma/2 \tag{9–58}$$

If light is propagated along [001], the birefringence manifests itself by two waves of index n_1 and n_2, respectively, for which the path retardation is

$$\delta_{\text{path}} = (n_2 - n_1)L = (q^{e}_{11} - q^{e}_{12}) n_0^3 \sigma L/2 \tag{9–59}$$

(where L is the length of path), or the phase difference

$$\Gamma = \pi n_0^3 (q^{e}_{11} - q^{e}_{12}) \sigma L/\lambda \tag{9–60}$$

Similarly, application of stress σ along the [111] axis will make the crystal uniaxial about that direction. Converting to the original axes gives, for the phase difference of light waves propagated perpendicular to the [111] direction,

$$\Gamma = \pi n_0^3 q_{44}^e \sigma L/\lambda \tag{9–61}$$

Experimental determination of the phase differences for these two stress directions can then give the photoelastic constants $q_{11}^e - q_{22}^e$ and q_{44}^e. To obtain q_{11}^e and q_{12}^e separately requires a direct measurement of the changes in refractive indices n_1 and n_2, for example by interferometric methods. A detailed discussion of various methods for measuring photoelastic and strain-optical constants, as well as the numerical values of these constants for a wide variety of crystals, is given by Narasimhamurty (1981).

Another much studied property of the $T(4)$ type is the piezoresistance Π, which is the change in the resistivity tensor produced by applied stress (Eq. (2–16) or Eq. (2–17)). The imposition of a stress (or strain) will change the electrical resistance of a crystal both because of changes in specific resistivity, ρ, and because of dimensional changes. In metals these two effects have the same order of magnitude, but in certain semiconductors the changes in ρ are far larger than the effects of dimensional changes. This can be interpreted in terms of the electronic wave functions of the semiconductors (Keyes, 1960). Because of this large effect, semiconductors such as Ge and Si have been shown to make very effective strain gauges (Mason, 1964; Thurston, 1964), with the additional advantage that they can be fabricated in very small sizes.

For upper cubic crystals there are three independent coefficients: Π_{11}, Π_{12} and Π_{44}. The constant $\Pi_{11} + 2\Pi_{12}$ is the pressure coefficient of piezoresistivity and the basis for pressure strain gauges. On the other hand, the quantity $\Pi_A = \Pi_{11} - \Pi_{12} - \Pi_{44}$ measures the anisotropy of the effect. It can be quite large for semiconductor materials, leading to a search for orientations that give extremal values (in general these are non-crystallographic orientations).

The property of magnetoresistivity is generally studied together with the Hall effect as the principal galvanomagnetic effects in metallic conductors. An early review of this subject is given by Jan (1957).

The electrostrictive tensor γ, Eq. (1–27), may be regarded as the dependence of the piezoelectric tensor **d** on electric field **E**, or

$$\varepsilon_{ij} = (d_{ijl}^0 + \gamma_{ijkl} E_k) E_l \tag{9–62}$$

where the parenthesized quantity is the piezoelectric tensor. In this way, all piezoelectric coefficients d_{ijl}^0, which are the values in the absence of an electric field, are modified by the field. But, more significantly, coefficients that were zero for zero field may now be present, though their values are small and proportional to the field components. In fact, crystals for which d^0_{ijl} is identically zero (e.g., crystals with inversion symmetry) may now show a non-zero piezoelectric effect proportional to the field. This type of second-order effect is sometimes referred to as a 'morphic effect'. It may be viewed as occurring due to the lowering of the crystal symmetry by the applied field. Thus the application of any field components that do not belong to the A_1 irrep must inevitably lower the symmetry of the crystal. In the higher-order effect the appropriate symmetry is that common to *both* the crystal and the field. For our purposes, the treatment of γ as a $T(4)$ tensor automatically takes care of these questions, giving us the allowed coefficients based on the symmetry of the crystal alone.

Other morphic effects have already been encountered, as, for example, in the case of the electro-optic effect, where crystals that are not normally birefringent can become so in the presence of an electric field. (See Section 7–1–3.) A similar remark applies to the photoelastic effect, Eq. (55), where a stress field is applied.

9–4 Relation between $T_S(2)$ and $T(2)$: magnetothermoelectric power

A property, such as the magnetothermoelectric power, Σ_{ijkl} in Eq. (2–28), relates a $T_S(2)$ tensor to a $T(2)$ tensor to give a 6×9 matter tensor of rank 4. Here we take **X** to be the $T_S(2)$ and **Y** the $T(2)$.

The $T(2)$ tensor has nine components which can be considered to form a nine-vector as already described in Chapter 7 (Eqs. (7–23)–(7–26)), in which the first six components form a symmetric tensor, and the last three form an antisymmetric tensor. Thus, **K** takes the block form

$$\mathbf{K} = \begin{array}{|c|} \hline 6 \times 6 \\ \\ \hline 3 \times 6 \\ \\ \hline \end{array}$$

in which the 6×6 block is a $T_S(2) \times T_S(2)$ tensor as given by Table 9–1, while the 3×6 block, representing $T(1)^{\text{ax}} \times T_S(2)$, is obtained from the first six columns of Table 7–2 which represent a $T(3)^{\text{ax}}$ tensor. These two blocks are easily combined; therefore, the matrices will not be presented separately.

It then remains to convert **K** to a four-index form $K_{ij} \rightarrow K_{klmn}$, where the scheme for converting j to mn is, as usual, $1 \rightarrow 11$, . . . and $4 \rightarrow 23$, ...

On the other hand, the conversion of i to kl follows Eqs. (7–26) and (7–27) (now for the variable Y rather than X). Thus, $i = 1, 2, 3$ becomes $kl = 11, 22$ and 33, respectively, while $(K_{4j} \pm K_{7j})/\sqrt{2} \rightarrow K_{23mn}$ and K_{32mn} for the case of plus and minus respectively.

Let us illustrate this conversion for the case of lower tetragonal crystals. From Tables 9–1 and 7–2 we obtain for K_{ij} (giving indices only)

$$\mathbf{K} = \begin{pmatrix} 11 & 12 & 13 & 0 & 0 & 16 \\ 12 & 11 & 13 & 0 & 0 & -16 \\ 31 & 31 & 33 & 0 & 0 & 0 \\ 0 & 0 & 0 & 44 & 45 & 0 \\ 0 & 0 & 0 & -45 & 44 & 0 \\ 61 & -61 & 0 & 0 & 0 & 66 \\ 0 & 0 & 0 & 74 & 75 & 0 \\ 0 & 0 & 0 & 75 & -74 & 0 \\ 91 & 91 & 93 & 0 & 0 & 0 \end{pmatrix} \tag{9–63}$$

where indices 74, 75, 91 and 93 have been renamed to suit the present numbering scheme. Making the conversions to four-index notation, we obtain the array

$$\begin{pmatrix} 1111 & 1122 & 1133 & 0 & 0 & 1112 \\ 1122 & 1111 & 1133 & 0 & 0 & -1112 \\ 3311 & 3311 & 3333 & 0 & 0 & 0 \\ 0 & 0 & 0 & 2323 & 2331 & 0 \\ 0 & 0 & 0 & -3231 & 3223 & 0 \\ 1211 & 1222 & 1233 & 0 & 0 & 1212 \\ 0 & 0 & 0 & 3223 & 3231 & 0 \\ 0 & 0 & 0 & -2331 & 2323 & 0 \\ -1222 & -1211 & -1233 & 0 & 0 & 1212 \end{pmatrix}$$

where

$$\left.\begin{aligned} &K_{1111} = K_{11},\ K_{1122} = K_{12},\ K_{1133} = K_{13},\ K_{1112} = K_{16}, \\ &K_{3311} = K_{31},\ K_{3333} = K_{33},\ K_{1211} = (K_{61} + K_{91})/\sqrt{2}, \\ &K_{1222} = -(K_{61} - K_{91})/\sqrt{2},\ K_{1212} = K_{66}/\sqrt{2},\ K_{1233} = K_{93}/\sqrt{2}, \\ &K_{2323} = (K_{44} + K_{74})/\sqrt{2},\ K_{2331} = (K_{45} + K_{75})/\sqrt{2}, \\ &K_{3231} = (K_{45} - K_{75})/\sqrt{2},\ K_{3223} = (K_{44} - K_{74})/\sqrt{2}. \end{aligned}\right\} \tag{9–64}$$

in terms of the non-zero two-index coefficients of Eq. (63). In carrying out this conversion, the reader should be aware of the array for the triclinic system (which involves no equalities due to symmetry) as given in Table 9–4.

Table 9–4. *Forms of a $T(4)$ tensor, such as magnetothermoelectric power, in four-index notation for various crystal symmetries (subscripts only)*

Triclinic

1111	1122	1133	1123	1131	1112
2211	2222	2233	2223	2231	2212
3311	3322	3333	3323	3331	3312
2311	2322	2333	2323	2331	2312
3111	3122	3133	3123	3131	3112
1211	1222	1233	1223	1231	1212
3211	3222	3233	3223	3231	3212
1311	1322	1333	1323	1331	1312
2111	2122	2133	2123	2131	2112

Monoclinic

1111	1122	1133	0	0	1112
2211	2222	2233	0	0	2212
3311	3322	3333	0	0	3312
0	0	0	2323	2331	0
0	0	0	3123	3131	0
1211	1222	1233	0	0	1212
0	0	0	3223	3231	0
0	0	0	1323	1331	0
2111	2122	2133	0	0	2112

Orthorhombic

1111	1122	1133	0	0	0
2211	2222	2233	0	0	0
3311	3322	3333	0	0	0
0	0	0	2323	0	0
0	0	0	0	3131	0
0	0	0	0	0	1212
0	0	0	3223	0	0
0	0	0	0	1331	0
0	0	0	0	0	2112

Lower tetragonal

1111	1122	1133	0	0	1112
1122	1111	1133	0	0	−1112
3311	3311	3333	0	0	0
0	0	0	2323	2331	0
0	0	0	−3231	3223	0
1211	1222	1233	0	0	1212
0	0	0	3223	3231	0
0	0	0	−2331	2323	0
−1222	−1211	−1233	0	0	1212

For *upper tetragonal*, set 1112, 1211, 1222, 1233, 2331, and 3231 equal to zero.

Table 9–4. *(cont.)*

Lower trigonal

1122 + 2(1212)	1122	1133	−2223	1131	$-\frac{1}{2}$(2111 + 1211)
1122	1122 + 2(1212)	1133	2223	−1131	$\frac{1}{2}$(2111 + 1211)
3311	3311	3333	0	0	0
−2322	2322	0	1331	−1323	−1311
3111	−3111	0	3123	3131	−3222
1211	−2111	1233	−1131	−2223	1212
−3222	3222	0	3131	−3123	−3111
1311	−1311	0	1323	1331	−2322
2111	−1211	−1233	−1131	−2223	1212

For *upper trigonal*, set 1131, 2111, 1211, 1323, 1311, 3111, 3123 and 1233 equal to zero.

Lower hexagonal

1122 + 2(1212)	1122	1133	0	0	0
1122	1122 + 2(1212)	1133	0	0	0
3311	3311	3333	0	0	0
0	0	0	1331	−1323	0
0	0	0	3123	3131	0
1211	−2111	1233	0	0	1212
0	0	0	3131	−3123	0
0	0	0	1323	1331	0
2111	−1211	−1233	0	0	1212

For *upper hexagonal*, set 1323, 3123, 1211, 2111 and 1233 equal to zero.

Lower cubic

1111	1122	1133	0	0	0
1133	1111	1122	0	0	0
1122	1133	1111	0	0	0
0	0	0	1212	0	0
0	0	0	0	1212	0
0	0	0	0	0	1212
0	0	0	1331	0	0
0	0	0	0	1331	0
0	0	0	0	0	1331

For *upper cubic*, set 1122 = 1133 and 1212 = 1331.

In a similar way, the **K** matrices of a property such as the magnetothermoelectric power can be constructed for crystals of all other symmetries, as presented in Table 9–4 in the four-index notation.

The reader may wish to consider the meaning of some of these four-index coefficients for the magnetothermoelectric power (see Problem 9–10).

9–5 Relation between $T(1)$ and $T_S(3)$: the second-order Hall effect

Consider the second-order Hall effect discussed in Section 2–4, which is the third-order effect of a magnetic B field on the resistivity ρ. It may be written as ρ_{ijklm}, a tensor of rank 5, but since it is antisymmetric in the indices i and j, it is better regarded as a $T(4)$ that relates an axial vector $T(1)^{\text{ax}}$ and a third-rank axial tensor $T_S(3)^{\text{ax}}$ that is totally symmetric in its three indices (since it comes from a triple product $B_jB_lB_m$). However, in view of the fact that both **Y** and **X** are axial, **K** is not; in fact, **K** has the same form as if we are relating a $T(1)$ to a $T_S(3)$, neither being axial. (This is a simpler problem to solve than that in which both tensors are axial.) Since the $T_S(3)$ tensor is totally symmetric in its three indices it has only ten independent components. In other words, we may solve the problem by taking **Y** to be a $T(1)$ vector and **X** to be a $T_S(3)$ tensor, giving a 3×10 matrix for **K**, with 30 components in the absence of crystal symmetry.

To deal with this problem we need the symmetry coordinates of a $T_S(3)$ tensor, or of the triple products of the set of coordinates x, y, z, just as a $T_S(2)$ tensor involved the double products of this set of coordinates. Such triple products have already been presented in the S-C-T tables (Appendix E) under the nomenclature $\beta_1, \ldots, \beta_{10}$, where the Weyl-normalized components of the $T_S(3)$ written as a ten-vector β are given by (see Eq. (4–5))

$$\left.\begin{array}{l} \beta_1 = x^3,\ \beta_2 = y^3,\ \beta_3 = z^3, \\ \beta_4 = \sqrt{3}x^2y,\ \beta_5 = \sqrt{3}x^2z,\ \beta_6 = \sqrt{3}y^2z,\ \beta_7 = \sqrt{3}xy^2 \\ \beta_8 = \sqrt{3}xz^2,\ \beta_9 = \sqrt{3}yz^2 \quad \text{and} \quad \beta_{10} = \sqrt{6}xyz \end{array}\right\} \qquad (9\text{–}65)$$

In three-index notation, subscripts 1–10 correspond to indices 111, 222, 333, 112, 113, 223, 221, 331, 332 and 123 respectively. The representation of the ten-vector $\boldsymbol{\beta}$ is denoted by Γ_β, and all its symmetry coordinates and their irreps are given in the S-C-T tables. Once we have these symmetry coordinates of $\boldsymbol{\beta}$, it is straightforward to calculate the non-zero components of K_{ij}. The coefficients obtained will involve those irreps in common between Γ_{xyz} and Γ_β.

We illustrate the procedure with a few examples, and present the results for all crystal classes in two-index form (K_{ij}) in Table 9–5 and in four-index form (K_{ijkl}) in Table 9–6.

Let us begin with the upper cubic classes (O_h, O, T_d). The symmetry coordinates of **Y** (which transforms as x, y, z) all belong to one of the T irreps, while the symmetry coordinates of **X** may be listed as $\bar{X}_1, \ldots, \bar{X}_{10}$ corresponding to X_1, X_2, X_3, $(X_4 + X_9)/\sqrt{2}$, $(X_4 - X_9)/\sqrt{2}$,

$(X_5+X_6)/\sqrt{2}$, $(X_5-X_6)/\sqrt{2}$, $(X_7+X_8)/\sqrt{2}$, $(X_7-X_8)/\sqrt{2}$, X_{10} respectively. (See the S-C-T tables for β.) Of these the two sets $(\bar{X}_1, \bar{X}_2, \bar{X}_3)$ and $(\bar{X}_8, \bar{X}_4, \bar{X}_6)$ belong to the same T irrep as x, y, z with similar orientations. Accordingly, the $\bar{\mathbf{K}}$ matrix, with subscripts only, takes the form

$$\bar{\mathbf{K}} = \begin{pmatrix} \mathrm{T},1 & 0 & 0 & 0 & 0 & 0 & 0 & \mathrm{T},2 & 0 & 0 \\ 0 & \mathrm{T},1 & 0 & \mathrm{T},2 & 0 & 0 & 0 & 0 & 0 & 0 \\ 0 & 0 & \mathrm{T},1 & 0 & 0 & \mathrm{T},2 & 0 & 0 & 0 & 0 \end{pmatrix} \qquad (9\text{–}66)$$

Finally, **K** takes the form given in Table 9–5, with

$$K_{11} = K_{\mathrm{T},1} \quad \text{and} \quad K_{17} = K_{\mathrm{T},2}/\sqrt{2}$$

In order to convert to four-index notation, we make use of Eq. (65) to note that

$$\left.\begin{aligned} &K_{ij} \rightarrow K_{ijjj} && \text{for } j = 1, 2, 3 \\ &K_{i4} \rightarrow \sqrt{3}K_{i112}, && \text{etc., for } j = 4\text{–}9 \\ &K_{i10} \rightarrow \sqrt{6}K_{i123} && \end{aligned}\right\} \qquad (9\text{–}67)$$

In this way, we obtain the result given in Table 9–6 with

$$K_{1111} = K_{11} \quad \text{and} \quad K_{1122} = K_{17}/\sqrt{3}$$

For the next illustration, consider the upper hexagonal classes (D_{6h}, D_6, C_{6v}, C_{3h}). Here $\Gamma_{xyz} = A_1 + E_1$ for group C_{6v} (and similarly for the others), while Γ_β contains β_3 and $\beta_5+\beta_6$ belonging to A_1, and the sets (β_9, β_8) and $(\beta_2+\beta_4/\sqrt{3},\ \beta_1+\beta_7/\sqrt{3})$ belonging to E_1 with the same orientation as (y, x). For the ten symmetry coordinates $\bar{\mathbf{X}}$, we may take

$$\begin{aligned} &(X_1+X_7/\sqrt{3})\sqrt{3}/2,\ (X_1-\sqrt{3}X_7)/2,\ (X_2+X_4/\sqrt{3})\sqrt{3}/2, \\ &\qquad (X_2-\sqrt{3}X_4)/2,\ X_3,\ (X_5+X_6)/\sqrt{2},\ (X_5-X_6)/\sqrt{2},\ X_8,\ X_9,\ X_{10} \end{aligned} \qquad (9\text{–}68)$$

The $\bar{\mathbf{K}}$ matrix then takes the form (subscripts only)

$$\bar{\mathbf{K}} = \begin{pmatrix} \mathrm{E}_1,2 & 0 & 0 & 0 & 0 & 0 & 0 & \mathrm{E}_1,1 & 0 & 0 \\ 0 & 0 & \mathrm{E}_1,2 & 0 & 0 & 0 & 0 & 0 & \mathrm{E}_1,1 & 0 \\ 0 & 0 & 0 & 0 & \mathrm{A}_1,1 & \mathrm{A}_1,2 & 0 & 0 & 0 & 0 \end{pmatrix} \qquad (9\text{–}69)$$

With the aid of the definitions of the $\bar{\mathbf{X}}$ coordinates, Eq. (68), it is straightforward to transform back to the unsymmetrized coordinates $\bar{\mathbf{X}}$ and to obtain the matrix **K** as given in Table 9–5, with

$$K_{18} = K_{\mathrm{E}_1,1}\ K_{17} = K_{\mathrm{E}_1,2}/2, \quad K_{33} = K_{\mathrm{A}_1,1}\ K_{35} = K_{\mathrm{A}_1,2}/\sqrt{2}$$

With the aid of Eq. (67), it is then straightforward to convert to the four-index form of **K** in Table 9–6.

Finally, we turn to the lower hexagonal classes (C_6, C_{3h} and C_{6h}). Here

Table 9–5. *Forms of a $T(4)$ tensor that relates a $T(1)$ to a $T_S(3)$, in two-index notation, for various crystal symmetries*

Triclinic

$$\begin{pmatrix} K_{11} & K_{12} & K_{13} & K_{14} & K_{15} & K_{16} & K_{17} & K_{18} & K_{19} & K_{110} \\ K_{21} & K_{22} & K_{23} & K_{24} & K_{25} & K_{26} & K_{27} & K_{28} & K_{29} & K_{210} \\ K_{31} & K_{32} & K_{33} & K_{34} & K_{35} & K_{36} & K_{37} & K_{38} & K_{39} & K_{310} \end{pmatrix}$$

Monoclinic

$$\begin{pmatrix} K_{11} & K_{12} & 0 & K_{14} & 0 & 0 & K_{17} & K_{18} & K_{19} & 0 \\ K_{21} & K_{22} & 0 & K_{24} & 0 & 0 & K_{27} & K_{28} & K_{29} & 0 \\ 0 & 0 & K_{33} & 0 & K_{35} & K_{36} & 0 & 0 & 0 & K_{310} \end{pmatrix}$$

Orthorhombic

$$\begin{pmatrix} K_{11} & 0 & 0 & 0 & 0 & 0 & K_{17} & K_{18} & 0 & 0 \\ 0 & K_{22} & 0 & K_{24} & 0 & 0 & 0 & 0 & K_{29} & 0 \\ 0 & 0 & K_{33} & 0 & K_{35} & K_{36} & 0 & 0 & 0 & 0 \end{pmatrix}$$

Lower trigonal

$$\begin{pmatrix} K_{11} & K_{12} & 0 & K_{12}/\sqrt{3} & K_{15} & -K_{15} & K_{11}/\sqrt{3} & K_{18} & K_{19} & \sqrt{2}K_{25} \\ -K_{12} & K_{11} & 0 & K_{11}/\sqrt{3} & K_{25} & -K_{25} & -K_{12}/\sqrt{3} & -K_{19} & K_{18} & -\sqrt{2}K_{15} \\ K_{31} & K_{32} & K_{33} & -\sqrt{3}K_{32} & K_{35} & K_{35} & -\sqrt{3}K_{31} & 0 & 0 & 0 \end{pmatrix}$$

Upper trigonal

$$\begin{pmatrix} K_{11} & 0 & 0 & 0 & 0 & 0 & K_{11}/\sqrt{3} & K_{18} & 0 & \sqrt{2}K_{25} \\ 0 & K_{11} & 0 & K_{11}/\sqrt{3} & K_{25} & -K_{25} & 0 & 0 & K_{18} & 0 \\ 0 & K_{32} & K_{33} & -\sqrt{3}K_{32} & K_{35} & K_{35} & 0 & 0 & 0 & 0 \end{pmatrix}$$

Lower tetragonal

$$\begin{pmatrix} K_{11} & K_{12} & 0 & K_{14} & 0 & 0 & K_{17} & K_{18} & K_{19} & 0 \\ -K_{12} & K_{11} & 0 & K_{17} & 0 & 0 & -K_{14} & -K_{19} & K_{18} & 0 \\ 0 & 0 & K_{33} & 0 & K_{35} & K_{35} & 0 & 0 & 0 & 0 \end{pmatrix}$$

Upper tetragonal

$$\begin{pmatrix} K_{11} & 0 & 0 & 0 & 0 & 0 & K_{17} & K_{18} & 0 & 0 \\ 0 & K_{11} & 0 & K_{17} & 0 & 0 & 0 & 0 & K_{18} & 0 \\ 0 & 0 & K_{33} & 0 & K_{35} & K_{35} & 0 & 0 & 0 & 0 \end{pmatrix}$$

Lower hexagonal

$$\begin{pmatrix} K_{11} & K_{12} & 0 & K_{12}/\sqrt{3} & 0 & 0 & K_{11}/\sqrt{3} & K_{18} & K_{19} & 0 \\ -K_{12} & K_{11} & 0 & K_{11}/\sqrt{3} & 0 & 0 & -K_{12}/\sqrt{3} & -K_{19} & K_{18} & 0 \\ 0 & 0 & K_{33} & 0 & K_{35} & K_{35} & 0 & 0 & 0 & 0 \end{pmatrix}$$

Upper hexagonal

$$\begin{pmatrix} K_{11} & 0 & 0 & 0 & 0 & 0 & K_{11}/\sqrt{3} & K_{18} & 0 & 0 \\ 0 & K_{11} & 0 & K_{11}/\sqrt{3} & 0 & 0 & 0 & 0 & K_{18} & 0 \\ 0 & 0 & K_{33} & 0 & K_{35} & K_{35} & 0 & 0 & 0 & 0 \end{pmatrix}$$

Lower cubic

$$\begin{pmatrix} K_{11} & 0 & 0 & 0 & 0 & 0 & K_{17} & K_{18} & 0 & 0 \\ 0 & K_{11} & 0 & K_{18} & 0 & 0 & 0 & 0 & K_{17} & 0 \\ 0 & 0 & K_{11} & 0 & K_{17} & K_{18} & 0 & 0 & 0 & 0 \end{pmatrix}$$

Table 9–5. *(cont.)*

Upper cubic

$$\begin{pmatrix} K_{11} & 0 & 0 & 0 & 0 & 0 & K_{17} & K_{17} & 0 & 0 \\ 0 & K_{11} & 0 & K_{17} & 0 & 0 & 0 & 0 & K_{17} & 0 \\ 0 & 0 & K_{11} & 0 & K_{17} & K_{17} & 0 & 0 & 0 & 0 \end{pmatrix}$$

$\Gamma_{xyz} = A + \tilde{E}_1$ for group C_6, with similar forms for the others, while Γ_β includes β_3 and $\beta_5 + \beta_6$ belonging to irrep A, while $\beta_8 \mp i\beta_9$ and $(\beta_1 + \beta_7/\sqrt{3}) \mp i(\beta_2 + \beta_4/\sqrt{3})$ belong to $\tilde{E}_1$. With the same set of symmetry coordinates $\bar{\mathbf{X}}$ as employed in the case of the upper hexagonals, we obtain, for the present case,

$$\bar{\mathbf{K}} = \begin{pmatrix} \tilde{E}_1,2^{re} & 0 & \tilde{E}_1,2^{im} & 0 & 0 & 0 & 0 & \tilde{E}_1,1^{re} & \tilde{E}_1,1^{im} & 0 \\ -\tilde{E}_1,2^{im} & 0 & \tilde{E}_1,2^{re} & 0 & 0 & 0 & 0 & -\tilde{E}_1,1^{im} & \tilde{E}_1,1^{re} & 0 \\ 0 & 0 & 0 & 0 & A_1,1 & A_1,1 & 0 & 0 & 0 & 0 \end{pmatrix} \quad (9\text{–}70)$$

The conversion to the **K** matrix of Table 9–5 then takes place similarly to the upper hexagonal case, as does the conversion to the four-index form in Table 9–6.

9–6 Other possibilities involving triple products

In the preceding section, we avoided the need to obtain triple products of axial vector components R_x, R_y, R_z by switching from a relation between $T(1)^{ax}$ and $T_S(3)^{ax}$ to one between $T(1)$ and $T_S(3)$, which gives the same form of **K**. In a similar way, note that a relation between $T(1)$ and $T_S(3)^{ax}$ gives the same form of **K** as one between $T(1)^{ax}$ and $T_S(3)$. In either case, **K** is a $T(4)^{ax}$, that is, an axial fourth-rank tensor that is symmetric in the last three indices. While there are no well-known properties of this type listed in Tables 1–1 and 2–1, it is clear that higher-order effects, involving either electric or magnetic fields, which fit this category do exist.

In order to obtain the forms of **K** for various crystal symmetries for such a tensor, making use of the symmetry coordinates of triple products β that we have already derived, we clearly want to handle this $T(4)$ case as a relation between a $T(1)^{ax}$ and a $T_S(3)$. It is then necessary to deal only with irreps that are common to both $\Gamma_\mathbf{R}$ and Γ_β. There will be many classes for which $\mathbf{K} \equiv 0$, particularly those having a center of symmetry. For the other classes, the form of **K** is obtained in a straightforward manner.

Table 9–6. *Forms of a $T(4)$ tensor that relates a $T(1)$ to a $T_S(3)$, in four-index notation, for various crystal symmetries (subscripts only for non-zero coefficients)*

Triclinic									
1111	1112	1113	1122	1123	1133	1222	1223	1233	1333
2111	2112	2113	2122	2123	2133	2222	2223	2233	2333
3111	3112	3113	3122	3123	3133	3222	3223	3233	3333
Monoclinic									
1111	1112	0	1122	0	1133	1222	0	1233	0
2111	2112	0	2122	0	2133	2222	0	2233	0
0	0	3113	0	3123	0	0	3223	0	3333
Orthorhombic									
1111	0	0	1122	0	1133	0	0	0	0
0	2112	0	0	0	0	2222	0	2233	0
0	0	3113	0	0	0	0	3223	0	3333
Lower trigonal									
1111	$\frac{1}{3}$(1222)	1113	$\frac{1}{3}$(1111)	1123	1133	1222	−1113	1233	0
−1222	$\frac{1}{3}$(1111)	1123	$-\frac{1}{3}$(1222)	−1113	−1233	1111	−1123	1133	0
3111	−3222	3113	−3111	0	0	3222	3113	0	3333
Upper trigonal									
1111	0	0	$\frac{1}{3}$(1111)	1123	1133	0	0	0	0
0	$\frac{1}{3}$(1111)	1123	0	0	0	1111	−1123	1133	0
0	−3222	3113	0	0	0	3222	3113	0	3333
Lower tetragonal									
1111	1112	0	1122	0	1133	1222	0	1233	0
−1222	1122	0	−1122	0	−1233	1111	0	1133	0
0	0	3113	0	0	0	0	3113	0	3333
Upper tetragonal									
1111	0	0	1122	0	1133	0	0	0	0
0	1122	0	0	0	0	1111	0	1133	0
0	0	3113	0	0	0	0	3113	0	3333
Lower hexagonal									
1111	$\frac{1}{3}$(1222)	0	$\frac{1}{3}$(1111)	0	1133	1222	0	1233	0
−1222	$\frac{1}{3}$(1111)	0	$-\frac{1}{3}$(1222)	0	−1233	1111	0	1133	0
0	0	3113	0	0	0	0	3113	0	3333
Upper hexagonal									
1111	0	0	$\frac{1}{3}$(1111)	0	1133	0	0	0	0
0	$\frac{1}{3}$(1111)	0	0	0	0	1111	0	1133	0
0	0	3113	0	0	0	0	3113	0	3333

Table 9–6. *(cont.)*

Lower cubic

1111	0	0	1122	0	1133	0	0	0	0
0	1133	0	0	0	0	1111	0	1122	0
0	0	1122	0	0	0	0	1133	0	1111

Upper cubic

1111	0	0	1122	0	1122	0	0	0	0
0	1122	0	0	0	0	1111	0	1122	0
0	0	1122	0	0	0	0	1122	0	1111

The same considerations may also be applied to tensors of rank 5. While none are explicitly listed in Tables 1–1 and 2–1, an important type of $T(5)$ will come from higher-order field effects on a $T(2)$ quantity. An example would be the relation of a $T_S(2)$ to a $T_S(3)$, where **K** is a $T(5)$ which is symmetric in the first two and in the last three indices. Then **K** will have components belonging only to irreps shared by both Γ_α and Γ_β. Thus, again, **K** will be identically zero for all classes having a center of symmetry, since, then Γ_α contains only irreps of the g-type and Γ_β contains those of the u-type. We will not work out tables of **K** for this case, since such properties are not of great interest, but it is important to note that we have the ability to develop such tables when needed with the aid of the available S-C-T tables.

Problems

9–1. Go through all the steps to obtain the form of **K** given in Table 9–1 for both upper and lower trigonal classes.

9–2. Verify Eqs. (29) and (30) for conversion of elastic constants to engineering notation.

9–3. Show that the reciprocal Young's modulus, s'_{33}, for a hexagonal crystal oriented such that $\cos(z', z) = \gamma_3$ is given by; $s'_{33} = (1 - \gamma_3^2)^2 s_{11} + \gamma_3^2(1 - \gamma_3^2)(2s_{13} + s_{44}) + \gamma_3^4 s_{33}$.

9–4. Show, by calculating s'_{14}, that for a cubic crystal in arbitrary orientation, torsion–flexure coupling occurs.

9–5. For wave propagation in a hexagonal crystal, verify Eqs. (42)–(46). Obtain similar expressions for a *tetragonal* crystal for v_l in [001], [100] and [110] directions and for v_t in direction [100] polarized first along [010], then along [001]; finally, for v_t in the [110] direction polarized along [1$\bar{1}$0].

9–6. Carry out the interconversion between elastic compliances s_{ij} and stiffnesses c_{ij} for a tetragonal crystal, making use of Eq. (11).

9–7. Show that, for a cubic crystal, the three group-theoretical elastic stiffness constants have the following significance: $c_A/3$ is equal to the bulk modulus, while c_E and c_T represent two different shear moduli. Draw diagrams showing the specific shear deformations involved in c_E and c_T.

9–8. Verify the conversion to engineering notation for the photoelastic tensor, Eq. (52), the strain-optical tensor, Eq. (53) and the electrostrictive tensor, Eq. (54).

9–9. Obtain the four-index matrix of Table 9–4, applicable to the magnetothermoelectric power, for upper hexagonal crystals.

9–10. Explain the physical meanings of four-index coefficients 1122, 1212, 1331 and 3333 that appear in Problem 9–9 for magnetothermoelectric power.

9–11. Do the same as Problem 9–9 for the lower trigonal classes.

9–12. Work out the $T(4)$ tensor relating a $T(1)$ and $T_S(3)$ as in Table 9–5 (in two-index form) and Table 9–6 (in four-index form) for the cases of upper and lower tetragonal classes.

9–13. Do the same for the upper trigonal classes.

9–14. After examining various possible higher-order effects, suggest a property whose matter tensor is a $T(4)^{ax}$ that is symmetric in the last three indices.

9–15. Do the same to obtain a $T(5)$ matter tensor that is symmetric in the first two and in the last three indices (i.e. a relation between a $T_S(2)$ and a $T_S(3)$).

10

Matter tensors of rank 6

The only matter tensor having a rank as high as 6 that appears in Chapters 1 and 2 is the so-called 'third-order elastic constant' tensor of type $T_S(6)$. (See Section 1–6 and Table 1–1.) This tensor couples a quantity **Y**, which is a thermodynamic tension, t_i, of type $T_S(2)$ to a quantity **X**, which is the symmetric product of Lagrangian strains, $\eta_i\eta_j$, of type $T_S(4)$. (Here, both t_i and η_j are second-rank symmetric tensors whose six components are written in single-index hypervector notation.) The resulting matter tensor **K** is then a $T_S(6)$ tensor, symmetric in the interchange of all the indices. The first objective of this chapter will be to obtain the independent components of such a $T_S(6)$ tensor. This will require us to go beyond the material contained in the S-C-T tables (Appendix E).

10–1 Relation between $T_S(2)$ and $T_S(4)$

We wish to consider the usual relation: $\mathbf{Y} = \mathbf{KX}$, in which **Y** is a $T_S(2)$ tensor that transforms as a six-vector, and **X** is a $T_S(4)$ tensor which transforms as the 21 symmetric products of six-vectors, $\alpha_i\alpha_j$. (Here we use the notation α for a $T_S(2)$ tensor as in the S-C-T tables and Eq. (4–4).) In terms of the single-index quantities **Y** and **X**, **K** then become a two-index (6 × 21) matrix.

The K_{ij} in this two-index form is then related to **K** in three-index form in which **X** is replaced by products, for example $\alpha_k\alpha_l$, through the Weyl normalization condition (Eq. (4–3)):

$$\begin{aligned} K_{ikl} &= K_{ij} \qquad (k = l \text{ or } j \leqslant 6) \\ &= K_{ij}/\sqrt{2} \quad (k \neq l \text{ or } j > 6) \end{aligned} \tag{10–1}$$

in which we are using a numbering scheme such that the first six values of j are those for which $k = l$.

Note that we are interested here in **K** tensors that are symmetric in the interchange of all three indices, representing the intrinsic symmetry of, for example, the third-order elastic constants. Such symmetry does not readily manifest itself in the two-index notation and, therefore, will have to be imposed as a separate condition. In any case, this intrinsic condition reduces the number of K_{ij} constants from 126 (i.e. 6×21) to only 56.

As usual, in order to consider the effects of crystal symmetry, it is necessary to transform both **Y** and **X** into symmetry coordinates. For **Y**, which is a $T_S(2)$, this is straightforward, with the aid of the S-C-T tables. We may represent the transformation into symmetry coordinates $\bar{\mathbf{Y}}$ as follows:

$$\bar{\mathbf{Y}} = \mathbf{S}\mathbf{Y} \tag{10–2}$$

The $T_S(4)$ tensor **X** may be said to transform as a 21-vector, that is, a set of 21 quantities δ_i which are the symmetric products of the $T_S(2)$ quantities α_k, employing the following numbering convention (and taking account of the Weyl normalization):

$$\left.\begin{array}{llll}
\delta_1 = \alpha_1^2 & \delta_7 = \sqrt{2}\alpha_1\alpha_2 & \delta_{13} = \sqrt{2}\alpha_2\alpha_4 & \delta_{19} = \sqrt{2}\alpha_4\alpha_5 \\
\delta_2 = \alpha_2^2 & \delta_8 = \sqrt{2}\alpha_1\alpha_3 & \delta_{14} = \sqrt{2}\alpha_2\alpha_5 & \delta_{20} = \sqrt{2}\alpha_4\alpha_6 \\
\delta_3 = \alpha_3^2 & \delta_9 = \sqrt{2}\alpha_1\alpha_4 & \delta_{15} = \sqrt{2}\alpha_2\alpha_6 & \delta_{21} = \sqrt{2}\alpha_5\alpha_6 \\
\delta_4 = \alpha_4^2 & \delta_{10} = \sqrt{2}\alpha_1\alpha_5 & \delta_{16} = \sqrt{2}\alpha_3\alpha_4 & \\
\delta_5 = \alpha_5^2 & \delta_{11} = \sqrt{2}\alpha_1\alpha_6 & \delta_{17} = \sqrt{2}\alpha_3\alpha_5 & \\
\delta_6 = \alpha_6^2 & \delta_{12} = \sqrt{2}\alpha_2\alpha_3 & \delta_{18} = \sqrt{2}\alpha_3\alpha_6 &
\end{array}\right\} \tag{10–3}$$

In order to take **X** (i.e. the set δ_i) into symmetry coordinates, it is advantageous to use an intermediate set of quantities which we call δ_i', expressed in terms of the products, not of the original α quantities, but of the symmetrized quantities $\bar{\alpha}_k$. We employ the same numbering scheme as in Eq. (3) to relate the 1-index δ_i' to the products of type $\bar{\alpha}_k\bar{\alpha}_l$, as follows:

$$\delta_1' = \bar{\alpha}_1^2; \quad \delta_7 = \sqrt{2}\bar{\alpha}_1\bar{\alpha}_2, \text{ etc.} \tag{10–4}$$

Since the $\bar{\alpha}$'s are already symmetrized, it is easier to convert the δ_i' quantities into symmetry coordinates that it would have been for the δ_i. This objective has been carried out in Table 10–1 for the various crystal symmetries. The table is given at the end of this chapter, page. 188. It is clear that, because all the $\bar{\alpha}$ quantities are the same for members of the same Laue group, the table for the δ_i' must also be the same. Thus, for example, the upper tetragonal groups may be represented by a selected member C_{4v}, and so on for the others. It should also be noted that for groups having inversion symmetry (g- and u-type irreps), the entire table of interest lies among the g-irreps, so that there is no need to present it explicitly.

In order to illustrate how Table 10–1 is obtained, we consider a product $\bar{\alpha}_i\bar{\alpha}_j$, and recall that if $\bar{\alpha}_i$ belongs to irrep γ_1 and $\bar{\alpha}_j$ to γ_2, then the product belongs to the direct product irrep $\gamma_1 \times \gamma_2$. (See Section 3–4.) In the case where γ_1 and γ_2 are both one-dimensional (including the $\tilde{\mathrm{E}}$ types), this uniquely defines the product irrep and the symmetry coordinate $\bar{\delta}_k$ that belongs to it. In cases where two- or three-dimensional irreps are involved, we need to obtain linear combinations of the δ_i' of the proper orientation, that is, matching those of the $\bar{\alpha}_k$ belonging to the same irrep. To accomplish this, we follow the procedures described in Section 3–5. (The reader will have the opportunity to derive some of the Tables 10–1 in the problems.)

In order to introduce the intrinsic symmetry, namely, that **K** in three-index form is totally symmetric with respect to all interchange of indices, we make use of our physical relationship in its intermediate form:

$$\bar{\mathbf{Y}} = \mathbf{K}'\mathbf{X}' \tag{10–5}$$

where the $\bar{\mathbf{Y}}$ are symmetrized, while the $\mathbf{X}'$ quantities are the same as the δ_i', that is, they are products of symmetrized α's of the type $\bar{\alpha}_k\bar{\alpha}_l$. The two-index and three-index K' quantities are then related exactly as in Eq. (1), namely,

$$\left.\begin{aligned} K'_{ikl} &= K'_{ij} \qquad (j \leqslant 6) \\ K'_{ikl} &= K'_{ij}/\sqrt{2} \quad (j > 6) \end{aligned}\right\} \tag{10–6}$$

The transformation between **K** and **K**$'$ is best given in three-index form, as

$$K'_{ijk} = S_{il}S^*_{jm}S^*_{kn}K_{lmn} \tag{10–7}$$

or

$$K_{ijk} = S^{\dagger}_{il}S^{\dagger *}_{jm}S^{\dagger *}_{kn}K'_{lmn} \tag{10–8}$$

in view of Eq. (2), realizing that $\bar{\alpha}$ obeys the same transformation, and that **S** is unitary. (The asterisks are needed to cover the cases of complex irreps.). It is noteworthy that, under such a unitary transformation **K**$'$ retains the symmetry with respect to the interchange of all indices, in the same way as in the case of a two-index **K** (see Appendix A, Section A–4). Thus, the requirement of intrinsic symmetry may be imposed on the K'_{ijk} quantities to generate the appropriate equalities of coefficients.

Finally, **X** may be converted to symmetry coordinates, $\bar{\mathbf{X}}$, using Table 10–1, to give the fully symmetrized relation

$$\bar{\mathbf{Y}} = \bar{\mathbf{K}}\bar{\mathbf{X}} \tag{10–9}$$

Here $\bar{Y}_i = Y^{(\gamma)}_{\text{dr}}$ and $\bar{X}_j = X^{(\gamma)}_{\text{ds}}$, while the matter tensor group-theoretical quantities become $\bar{K}_{ij} = K_{\gamma,\text{rs}}$ with the non-zero coefficients being determined by the Fundamental Theorem. In the above, the subscript d represents a degeneracy index while r and s are repeat indices, in the usual way.

It is now appropriate to review the procedure that we shall follow in determining **K** for various crystal symmetries.

1. We begin with Eq. (9) in symmetrized form, relating the symmetry coordinates of $\bar{\mathbf{Y}}$ and $\bar{\mathbf{X}}$ to obtain the quantities $K_{\gamma,\text{rs}}$ by employing the Fundamental Theorem. Note that the present case differs from all previous ones where K components belonging to different irreps were independent of each other. Here, however, there are equalities due to the intrinsic symmetry of the K'_{ijk} with respect to interchange of the indices i, j and k. These equalities can be handled group theoretically by making use of direct products of the irreps γ_1, γ_2 and γ_3 to which $\bar{\text{Y}}$, $\bar{\alpha}_i$ and $\bar{\alpha}_j$ belong. However, the reader will probably find it more convenient to wait until step (2) to apply this criterion.
2. Convert $\bar{\mathbf{K}}$ into the two-index K'_{ij} using Table 10–1, and then to K'_{ikl} using Eq. (6). It is then easy to apply the conditions of intrinsic symmetry to K'_{ikl}. This results in equalities among the group-theoretical coefficients obtained in step (1), with a consequent reduction in the number of independent coefficients.
3. Finally, we convert the three-index K'_{lmn} coefficients to three-index K_{ijk}, using Eq. (8).

Note that, following this method, there is no need to make use of **K** in two-index notation.

Because of the large number (56) of matter tensor components before crystal symmetry is introduced, the study of such $T_S(6)$ tensors, and specifically of the third-order elastic constants, has been of interest only for relatively high-symmetry crystals. Accordingly, we focus on the higher symmetries in working out these results, but present tables that include all symmetries.

10–1–1 Case of the upper hexagonal groups

Even though the upper cubics have the highest symmetry and the smallest number of independent constants, let us begin this time with consideration of the upper hexagonal groups, of which C_{6v} can be considered typical. This case is also of special interest because of past difficulties in solving the problem correctly (Hearmon, 1953).

In the case of C_{6v}, we readily see from Table 10–1 that $\Gamma_\alpha = 2A_1 + E_1 + E_2$, while $\Gamma_\delta = 5A_1 + B_1 + B_2 + 3E_1 + 4E_2$. Thus from the

Fundamental Theorem, one can see immediately that there are 17 constants before taking into account the intrinsic symmetry. These constants are

$A_1,11$	$A_1,21$	$E_1,11$	$E_2,11$
$A_1,12$	$A_1,22$	$E_1,12$	$E_2,12$
$A_1,13$	$A_1,23$	$E_1,13$	$E_2,13$
$A_1,14$	$A_1,24$		$E_2,14$
$A_1,15$	$A_1,25$		

where we have listed only the subscripts of $K_{\gamma,rs}$.

The next step is to relate the above group-theoretical constants to K'_{ij} and then to the three-index K'_{ikl}. With the aid of Table 10–1, this is readily accomplished. For the A_1 coefficients the following results are obtained:

$$\left.\begin{aligned} A_1,11 &= K'_{11} = K'_{111} \\ A_1,12 &= K'_{12} = K'_{122} \\ A_1,13 &= K'_{17} = \sqrt{2}K'_{112} \\ A_1,21 &= K'_{21} = K'_{211} \\ A_1,22 &= K'_{22} = K'_{222} \\ A_1,23 &= K'_{27} = \sqrt{2}K'_{212} \end{aligned}\right\} \qquad (10\text{–}10)$$

while

$$\left.\begin{aligned} A_1,14 &= (K'_{13} + K'_{14})/\sqrt{2} = (K'_{133} + K'_{144})/\sqrt{2} \\ A_1,15 &= (K'_{15} + K'_{16})/\sqrt{2} = (K'_{155} + K'_{166})/\sqrt{2} \\ A_1,24 &= (K'_{23} + K'_{24})/\sqrt{2} = (K'_{233} + K'_{244})/\sqrt{2} \\ A_1,25 &= (K'_{25} + K'_{26})/\sqrt{2} = (K'_{255} + K'_{266})/\sqrt{2} \end{aligned}\right\} \qquad (10\text{–}11)$$

From the intrinsic symmetry of the these K'_{ikl} coefficients, we find

$$A_1,12 = (1/\sqrt{2})A_1,23 = K'_{122} \qquad (10\text{–}12)$$

and

$$(1/\sqrt{2})A_1,13 = A_1,21 = K'_{112} \qquad (10\text{–}13)$$

For the E_1 coefficients, the following results are obtained:

$$\left.\begin{aligned} E_1,\ 11 &= K'_{38} = K'_{49} = \sqrt{2}K'_{313} = \sqrt{2}K'_{414} \\ E_1,\ 12 &= K'_{3\underline{12}} = K'_{4\underline{13}} = \sqrt{2}K'_{323} = \sqrt{2}K'_{424} \\ E_1,\ 13 &= (K'_{3\underline{20}} - K'_{3\underline{17}})/\sqrt{2} = (K'_{4\underline{18}} + K'_{4\underline{19}})/\sqrt{2} \\ &= K'_{346} - K'_{335} = K'_{436} + K'_{445} \end{aligned}\right\} \qquad (10\text{–}14)$$

(Here each underlined pair of digits is to be read as a *single* number of value greater than 9. The underlining is done to avoid confusion with the three-index quantities.) Equation (14) provides equalities that follow from the second part of the Fundamental Theorem (Section 4–4), related to two-dimensional irreps. Specifically,

$$\left.\begin{aligned} K'_{133} &= K'_{144} \\ K'_{233} &= K'_{244} \\ K'_{335} &= -K'_{445} \end{aligned}\right\} \qquad (10\text{–}15)$$

In addition, comparison of Eq. (14) with Eq. (11) yields two additional relations that follow from the intrinsic symmetry:

$$\mathrm{A}_1, 14 = \mathrm{E}_1, 11 = \sqrt{2}K'_{133} \qquad (10\text{–}16)$$

$$\mathrm{A}_1, 24 = \mathrm{E}_1, 12 = \sqrt{2}K'_{233} \qquad (10\text{–}17)$$

Finally, we examine the E_2 coefficients, and obtain:

$$\left.\begin{aligned} \mathrm{E}_2, 11 &= K'_{5\underline{10}} = K'_{6\underline{11}} = \sqrt{2}K'_{515} = \sqrt{2}K'_{616} \\ \mathrm{E}_2, 12 &= K'_{5\underline{14}} = K'_{6\underline{15}} = \sqrt{2}K'_{525} = \sqrt{2}K'_{626} \\ \mathrm{E}_2, 13 &= (K'_{54} - K'_{53})/\sqrt{2} = K'_{6\underline{16}} = (K'_{544} - K'_{533})/\sqrt{2} = \sqrt{2}K'_{634} \\ \mathrm{E}_2, 14 &= (K'_{55} - K'_{56})/\sqrt{2} = -K'_{6\underline{21}} = (K'_{555} - K'_{566})/\sqrt{2} = -\sqrt{2}K'_{656} \end{aligned}\right\} \qquad (10\text{–}18)$$

These results give the additional Fundamental-Theorem equalities:

$$\left.\begin{aligned} K'_{155} &= K'_{166} \\ K'_{255} &= K'_{266} \\ K'_{335} &= -K'_{346} \\ K'_{555} &= -K'_{566} \end{aligned}\right\} \qquad (10\text{–}19)$$

Comparison of the results of Eq. (18) with Eqs. (11) and (14) gives three additional intrinsic equalities:

$$\mathrm{A}_1, 15 = \mathrm{E}_2, 11 = \sqrt{2}K'_{155} \qquad (10\text{–}20)$$

$$\mathrm{A}_1, 25 = \mathrm{E}_2, 12 = \sqrt{2}K'_{255} \qquad (10\text{–}21)$$

$$\mathrm{E}_1, 13 = \sqrt{2}\mathrm{E}_2, 13 = -2K'_{335} \qquad (10\text{–}22)$$

In all, we have added seven equalities: Eqs. (12), (13), (16), (17), (20), (21) and (22), as a consequence of the intrinsic symmetry of the K'_{ikl}. Accordingly, the total number of independent coefficients is reduced from 17 to 10.

The final step is to convert the ten-index K' coefficients to the three-index K coefficients. For this we make use of Eq. (8), obtaining the $\mathbf{S}^{\dagger}$

matrix from the first column of Table 10–1, namely,

$$\mathbf{S}^\dagger = \begin{pmatrix} 1/\sqrt{2} & 0 & 0 & 0 & 1/\sqrt{2} & 0 \\ 1/\sqrt{2} & 0 & 0 & 0 & -1/\sqrt{2} & 0 \\ 0 & 1 & 0 & 0 & 0 & 0 \\ 0 & 0 & 1 & 0 & 0 & 0 \\ 0 & 0 & 0 & 1 & 0 & 0 \\ 0 & 0 & 0 & 0 & 0 & 1 \end{pmatrix} \qquad (10\text{–}23)$$

In carrying out the conversion from K' to the K coefficients, we must also use the equalities among the K'_{ikl} coefficients that came from the second part of the Fundamental Theorem, namely, Eqs. (15) and (19) in the present case. The conversion is straightforward, though a bit tedious and yields the following results:

$$K_{111} = (1/\sqrt{2})^3(K'_{111} + K'_{555} + 3K'_{155}) \qquad (10\text{–}24)$$

$$K_{222} = (1/\sqrt{2})^3(K'_{111} - K'_{555} + 3K'_{155}) \qquad (10\text{–}25)$$

$$K_{112} = (1/\sqrt{2})^3(K'_{111} - K'_{555} - K'_{155}) \qquad (10\text{–}26)$$

$$K_{122} = (1/\sqrt{2})^3(K'_{111} + K'_{555} - K'_{155}) \qquad (10\text{–}27)$$

The last of these four equations is clearly not independent, but may be related to the others by

$$K_{122} = K_{111} - K_{222} + K_{112} \qquad (10\text{–}28)$$

In a similar way, we obtain other independent coefficients:

$$K_{113} = K_{223} = (1/2)(K'_{112} + K'_{255}) \qquad (10\text{–}29)$$

$$K_{123} = (1/2)(K'_{112} - K'_{255}) \qquad (10\text{–}30)$$

$$K_{133} = K_{233} = (1/\sqrt{2})K'_{122} \qquad (10\text{–}31)$$

$$K_{144} = K_{255} = (1/\sqrt{2})(K'_{133} + K'_{335}) \qquad (10\text{–}32)$$

$$K_{155} = K_{244} = (1/\sqrt{2})(K'_{133} - K'_{355}) \qquad (10\text{–}33)$$

$$K_{333} = K'_{222} \qquad (10\text{–}34)$$

$$K_{344} = K_{355} = K'_{233} \qquad (10\text{–}35)$$

and those that are non-zero but expressible in terms of other K_{ikl} coefficients:

$$2K_{166} = -2K_{111} - K_{112} + 3K_{222} \qquad (10\text{–}36)$$

$$2K_{266} = 2K_{111} - K_{112} - K_{222} \qquad (10\text{–}37)$$

$$K_{366} = K_{113} - K_{123} \qquad (10\text{–}38)$$

$$K_{456} = (1/\sqrt{2})(K_{155} - K_{144}) \qquad (10\text{–}39)$$

All these results are listed in Table 10–2, which gives the independent

Table 10–2. *Listing of the non-zero three-index coefficients of a* $T(6)$ *tensor relating a* $T_S(2)$ *to a* $T_S(4)$*, for the various Laue groups as well as isotropic symmetry (after Brugger, 1965). Footnotes a through n give relations among the coefficients.*

The Laue groups are as follows: N (triclinic), M (monoclinic), O (orthorhombic), T (tetragonal), C (cubic), R (trigonal), H (hexagonal), I (isotropic), while I and II refer to upper and lower symmetries, respectively

N	M	O	TII	TI	CII	CI	RII	RI	HII	HI	I
111	111	111	111	111	111	111	111	111	111	111	111
112	112	112	112	112	112	112	112	112	112	112	112
113	113	113	113	113	113	112	113	113	113	113	112
114	0	0	0	0	0	0	114	114	0	0	0
115	115	0	0	0	0	0	115	0	0	0	0
116	0	0	116	0	0	0	116	0	116	0	0
122	122	122	112	112	113	112	122[a]	122[a]	122[a]	122[a]	112
123	123	123	123	123	123	123	123	123	123	123	123
124	0	0	0	0	0	0	124	124	0	0	0
125	125	0	0	0	0	0	125	0	0	0	0
126	0	0	0	0	0	0	−116	0	−116	0	0
133	133	133	133	133	112	112	133	133	133	133	112
134	0	0	0	0	0	0	134	134	0	0	0
135	135	0	0	0	0	0	135	0	0	0	0
136	0	0	136	0	0	0	0	0	0	0	0
144	144	144	144	144	144	144	144	144	144	144	144[l]
145	0	0	145	0	0	0	145	0	145	0	0
146	146	0	0	0	0	0	146[b]	0	0	0	0
155	155	155	155	155	155	155	155	155	155	155	155[m]
156	0	0	0	0	0	0	156[c]	156[c]	0	0	0
166	166	166	166	166	166	155	166[d]	166[d]	166[d]	166[d]	155[m]
222	222	222	111	111	111	111	222	222	222	222	111
223	223	223	113	113	112	112	113	113	113	113	112
224	0	0	0	0	0	0	224[e]	224[e]	0	0	0
225	225	0	0	0	0	0	225[f]	0	0	0	0
226	0	0	−116	0	0	0	116	0	116	0	0
233	233	233	133	133	113	112	133	133	133	133	112
234	0	0	0	0	0	0	−134	−134	0	0	0
235	235	0	0	0	0	0	−135	0	0	0	0
236	0	0	−136	0	0	0	0	0	0	0	0
244	244	244	155	155	166	155	155	155	155	155	155[m]
245	0	0	−145	0	0	0	−145	0	−145	0	0
246	246	0	0	0	0	0	246[g]	0	0	0	0
255	255	255	144	144	144	144	144	144	144	144	144[l]
256	0	0	0	0	0	0	256[h]	256[h]	0	0	0
266	266	266	166	166	155	155	266[i]	266[i]	266[i]	266[i]	155[m]
333	333	333	333	333	111	111	333	333	333	333	111
334	0	0	0	0	0	0	0	0	0	0	0
335	335	0	0	0	0	0	0	0	0	0	0

Table 10–2. *(cont.)*

N	M	O	TII	TI	CII	CI	RII	RI	HII	HI	I
336	0	0	0	0	0	0	0	0	0	0	0
344	344	344	344	344	155	155	344	344	344	344	155[m]
345	0	0	0	0	0	0	0	0	0	0	0
346	346	0	0	0	0	0	−135	0	0	0	0
355	355	355	344	344	166	155	344	344	344	344	155[m]
356	0	0	0	0	0	0	134	134	0	0	0
366	366	366	366	366	144	144	366[j]	366[j]	366[j]	366[j]	144[l]
444	0	0	0	0	0	0	444	444	0	0	0
445	445	0	0	0	0	0	445	0	0	0	0
446	0	0	446	0	0	0	145	0	145	0	0
455	0	0	0	0	0	0	−444	−444	0	0	0
456	456	456	456	456	456	456	456[k]	456[k]	456[k]	456[k]	456[n]
466	0	0	0	0	0	0	124	124	0	0	0
555	555	0	0	0	0	0	−445	0	0	0	0
556	0	0	−446	0	0	0	−145	0	−145	0	0
566	566	0	0	0	0	0	125	0	0	0	0
666	0	0	0	0	0	0	−116	0	−116	0	0

[a] $K_{122} = K_{111} + K_{112} - K_{222}$.
[b] $K_{146} = -(1/\sqrt{2})(K_{115} + 3K_{125})$.
[c] $K_{156} = (1/\sqrt{2})(K_{114} + 3K_{124})$.
[d] $K_{166} = \frac{1}{2}(-2K_{111} - K_{112} + 3K_{222})$.
[e] $K_{224} = -K_{114} - 2K_{124}$.
[f] $K_{225} = -K_{115} - 2K_{125}$.
[g] $K_{246} = (1/\sqrt{2})(-K_{115} + K_{125})$.
[h] $K_{256} = (1/\sqrt{2})(K_{114} - K_{124})$.
[i] $K_{266} = \frac{1}{2}(2K_{111} - K_{112} - K_{222})$.
[j] $K_{366} = K_{113} - K_{123}$.
[k] $K_{456} = (1/\sqrt{2})(-K_{144} + K_{155})$.
[l] $K_{144} = K_{112} - K_{123}$.
[m] $K_{155} = \frac{1}{2}(K_{111} - K_{112})$.
[n] $K_{456} = (1/2\sqrt{2})(K_{111} - 3K_{112} + 2K_{123})$.

non-zero coefficients K_{ikl} for all the Laue groups. The table gives only the three indices, and all dependence relationships of the types of Eqs. (28) and (36)–(39) are given as footnotes.

10–1–2 Case of the upper cubic and isotropic materials

We turn now to the upper cubics of which the group O is representative. Table 10–1 shows that $\Gamma_\alpha = A_1 + E + T_2$, while $\Gamma_\delta = 3A_1 + 3E + T_1 + 3T_2$. We, therefore, see immediately that there are nine independent constants before intrinsic symmetry is introduced. These group-theoretical constants are (subscripts only)

$$\begin{matrix} \mathrm{A_1,11} & \mathrm{E,11} & \mathrm{T_2,11} \\ \mathrm{A_1,12} & \mathrm{E,12} & \mathrm{T_2,12} \\ \mathrm{A_1,13} & \mathrm{E,13} & \mathrm{T_2,13} \end{matrix}$$

Table 10–1 allows us to relate these constants to the two-index K'_{ij} and then to the three-index K'_{ikl} coefficients, as follows.

$$\mathrm{A_1,11} = K'_{11} = K'_{111} \tag{10–40}$$

$$\mathrm{A_1,12} = (1/\sqrt{2})(K'_{12} + K'_{13}) = (1/\sqrt{2})(K'_{122} + K'_{133}) \tag{10–41}$$

$$\mathrm{A_1,13} = (1/\sqrt{3})(K'_{14} + K'_{15} + K'_{16}) = (1/\sqrt{3})(K'_{144} + K'_{155} + K'_{166}) \tag{10–42}$$

Similarly, for the E coefficients:

$$\mathrm{E,11} = K'_{27} = K'_{38} = \sqrt{2}K'_{212} = \sqrt{2}K'_{313} \tag{10–43}$$

$$\mathrm{E,12} = (1/\sqrt{2})(K'_{22} - K'_{23}) = -K'_{3\underline{12}} = (1/\sqrt{2})(K'_{222} - K'_{233}) = \sqrt{2}K'_{323} \tag{10–44}$$

$$\begin{aligned} \mathrm{E,13} &= (1/\sqrt{6})(2K'_{24} - K'_{25} - K'_{26}) = (1/\sqrt{2})(K'_{35} - K'_{36}) \\ &= (1/\sqrt{6})(2K'_{244} - K'_{255} - K'_{266}) = (1/\sqrt{2})(K'_{355} - K'_{366}) \end{aligned} \tag{10–45}$$

and, for the T_2 coefficients:

$$\mathrm{T_2,11} = K'_{49} = K'_{5\underline{10}} = K'_{6\underline{11}} = \sqrt{2}K'_{414} = \sqrt{2}K'_{515} = \sqrt{2}K'_{616} \tag{10–46}$$

$$\begin{aligned} \mathrm{T_2,12} &= -K'_{4\underline{13}} = (1/2)(K'_{5\underline{14}} - \sqrt{3}K'_{5\underline{17}}) + (1/2)(K'_{6\underline{15}} + \sqrt{3}K'_{6\underline{18}}) \\ &= -\sqrt{2}K'_{424} + (1/\sqrt{2})(K'_{525} - \sqrt{3}K'_{535}) \\ &= (1/\sqrt{2})(K'_{626} + \sqrt{3}K'_{636}) \end{aligned} \tag{10–47}$$

$$\mathrm{T_2,13} = K'_{4\underline{21}} = K'_{5\underline{20}} = K'_{6\underline{19}} = \sqrt{2}K'_{456} = \sqrt{2}K'_{546} = \sqrt{2}K'_{645} \tag{10–48}$$

Making use of these relations and the intrinsic symmetry of the K'_{ikl} coefficients, we obtain

$$\mathrm{A_1,11} = K'_{111} \tag{10–49}$$

$$\mathrm{A_1 12} = \mathrm{E,11} = \sqrt{2}K'_{122} \tag{10–50}$$

with

$$K'_{122} = K'_{133} \tag{10–51}$$

$$\mathrm{A_1,13} = \sqrt{(3/2)}\mathrm{T_2,11} = \sqrt{3}K'_{144} \tag{10–52}$$

with

$$K'_{144} = K'_{155} = K'_{166} \tag{10–53}$$

$$\mathrm{E,12} = \sqrt{2}K'_{222} \tag{10–54}$$

with

$$K'_{222} = -K'_{233} \tag{10–55}$$

$$E,13 = (\sqrt{3}/2)T_2,12 = \sqrt{(3/2)}K'_{244} \tag{10–56}$$

with

$$K'_{244} = -2K'_{255} = -2K'_{266} = (2/\sqrt{3})K'_{355} = -(2/\sqrt{3})K'_{366} \tag{10–57}$$

and

$$T_2,13 = \sqrt{2}K'_{456} \tag{10–58}$$

The three equalities of the group-theoretical coefficients Eqs. (50), (52) and (56), due to the intrinsic symmetry of $\mathbf{K}'$, reduce the number of independent constants from nine to only six.

Finally we need to convert the K'_{ikl} to the unprimed K_{ijk} coefficients, using Eq. (8) and the $\mathbf{S}^\dagger$ matrix:

$$\mathbf{S}^\dagger = \begin{pmatrix} 1/\sqrt{3} & 2/\sqrt{6} & 0 & 0 & 0 & 0 \\ 1/\sqrt{3} & -1/\sqrt{6} & 1/\sqrt{2} & 0 & 0 & 0 \\ 1/\sqrt{3} & -1/\sqrt{6} & -1/\sqrt{2} & 0 & 0 & 0 \\ 0 & 0 & 0 & 1 & 0 & 0 \\ 0 & 0 & 0 & 0 & 1 & 0 \\ 0 & 0 & 0 & 0 & 0 & 1 \end{pmatrix} \tag{10–59}$$

The results for the independent K_{ijk} coefficients are

$$K_{111} = (1/\sqrt{3})^3(K'_{111} + 2\sqrt{2}K'_{222} + 6K'_{122} \tag{10–60}$$

$$K_{112} = (1/\sqrt{3})^3(K'_{111} - \sqrt{2}K'_{222}) \tag{10–61}$$

$$K_{123} = (1/\sqrt{3})^2(K'_{111} + 2\sqrt{2}K'_{222} - 3K'_{122}) \tag{10–62}$$

$$K_{144} = (1/\sqrt{3})(K'_{144} + \sqrt{2}K'_{244}) \tag{10–63}$$

$$K_{155} = (1/\sqrt{3})[K'_{144} - (\sqrt{2}/2)K'_{244}] \tag{10–64}$$

$$K_{456} = K'_{456} \tag{10–65}$$

These results, together with equalities of the types $K_{222} = K_{333} = K_{111}$, $K_{155} = K_{166}$ and quite a few others given in Table 10–2, complete the solution of the problem for the upper cubic crystals.

The case of *isotropic media* is an important one for the study of third-order elastic constants. Just as for the case of second-order elastic constants, total isotropy means an additional reduction in the number of constants from the upper cubic symmetry. Rather than attempting to solve the isotropic case by using the full-rotational symmetry group R(3), however, it is much easier to make use of what we have already done. Specifically, it is easy to see that complete rotational isotropy can be obtained by combining cubic symmetry (O group) with full rotational symmetry $C_{\infty v}$ about any single axis, say the z-axis. (The cubic symmetry, of course, allows for the equivalence of three perpendicular axes.) But it may be seen from the S-C-T tables for the quantities α_i that the hexagonal C_{6v} gives results equivalent to $C_{\infty v}$. It may be concluded that *the* **K** *tensor for the isotropic case must have the combined symmetry of the upper cubic and upper hexagonal groups*. Examination of Table 10–2 then shows that the isotropic result is the same as that for upper cubic with three additional equalities:

$$(1/2)K_{366} = (1/2)K_{144} = 2K_{112} - K_{123} \qquad (10\text{–}66)$$

$$(1/2)K_{166} = (1/2)K_{155} = 3K_{111} - K_{112} \qquad (10\text{–}67)$$

$$(1/2\sqrt{3})K_{456} = 6K_{111} - 6K_{112} + 2K_{123} \qquad (10\text{–}68)$$

where we have utilized the equalities $K_{222} = K_{111}$ and $K_{113} = K_{112}$ that apply to the cubic case. In view of these equations, the number of independent constants is reduced from six to only three. (The reader may wish to apply this same method to some of the $T(4)$ tensors of Chapter 9.)

Having completed these cases of highest symmetry, we leave the others, covered by Table 10–2, to problems for the reader. Note, however, that for symmetries below the uniaxial classes, the $\bar{\alpha}$ and α quantities become the same. This serves to make the problem simpler for those cases, even though the number of independent constants increases greatly.

10–2 Application to third-order elastic constants

The literature on third-order elastic constants (TOECs) always quotes values of the three-index engineering constants C^{e}_{ijk} (as in Eq. (1–39)) which are related to the three-index quantities of Tables 10–2 by

$$\begin{aligned} C^{e}_{ijk} &= C_{ijk} && (i, j, k \text{ all} \leqslant 3) \\ &= (1/\sqrt{2})C_{ijk} && (\text{one index} \geqslant 4) \\ &= (1/2)C_{ijk} && (\text{two indices} \geqslant 4) \\ &= (1/2\sqrt{2})C_{ijk} && (\text{all indices} \geqslant 4) \end{aligned} \qquad (10\text{–}69)$$

Let us take a moment to discuss the importance of TOECs. These terms represent the lowest form of *anharmonic interactions* in crystals. In the absence of TOECs, so that the lattice potential energy involves quadratic terms only (see Eq. (1–39)), the stress–strain relations would be linear and the elastic constants c_{ij} independent of the strains. The consequences of such simple relationships are that lattice waves would pass each other without interaction or attenuation, that is, they only superimpose but do not change form with time. A solid that obeys such a model would be a one for which:

(a) the thermal conductivity is infinite (or thermal resistivity = 0);
(b) there is no thermal expansion;
(c) the c_{ij} are independent of pressure and temperature.

In terms of the 'phonon' concept (quantized lattice vibrations), each having a wave vector **k** and angular frequency ω, there can be no phonon–phonon interaction without TOECs (Burns, 1985). Processes such as thermal conductivity depend on phonon–phonon interaction processes in which the equations for 'conservation' of **k** and ω take the forms

$$\mathbf{k}_1 + \mathbf{k}_2 = \mathbf{k}_3 + \mathbf{G} \tag{10–70}$$

$$\omega_1 + \omega_2 = \omega_3 \tag{10–71}$$

Here the subscripts 1 and 2 represent two incident phonons and 3 represents an emerging phonon, while **G** is a reciprocal-lattice vector. Such three-phonon processes require at least cubic terms in the lattice potential energy, that is, the presence of TOECs. Physically, the phenomenon can be seen as follows. One lattice wave causes a periodic elastic strain which, because of anharmonic interactions, modulates the c_{ij} coefficients in space and time. The second wave perceives this modulation and is scattered, as from a moving three-dimensional grating.

Because of the large number of TOECs, they have been studied primarily in high-symmetry crystals and in isotropic materials.

In considering the methods of obtaining TOECs, we first recall that ultrasonic wave propagation offers a high-precision method for obtaining the c_{ij} coefficients. (See Section 9–2.) Accordingly, the most straightforward method of determining TOECs is to carry out ultrasonic velocity measurements under the influence of static applied strains. In such a case, the ultrasonic velocities give the *adiabatic* constants c_{ij}, while the application of constant strain is *isothermal*. Such measurements, therefore, give mixed adiabatic–isothermal TOECs (Wallace, 1970). The difference between adiabatic and isothermal moduli is small, however, and corrections can readily be made when required (see Section 1–5). This method has been used by many workers (Thurston, 1964; Brugger, 1965). One of the difficulties is that, in order to obtain high precision for the TOECs, it is desirable to use static strains that are as large as possible. On the other hand, strains that have a shear component can give rise to plastic deformation, and must be kept small enough to avoid this problem. No such problem exists when hydrostatic pressure is applied, however; such stress (and, therefore, strain) can be made as large as is possible without producing plastic deformation. It is, therefore, easy to study those TOECs that are given by the effect of pressure (and, therefore, of change of volume) on the c_{ij} coefficients. Of course, such measurements do not yield all the TOECs.

For the upper cubic case, application of hydrostatic pressure means that $\eta_1 = \eta_2 = \eta_3 = \eta$. To obtain the appropriate TOECs, we recognize that, in symmetry coordinates, this static strain is of the type η_A $(= \sqrt{3}\eta)$. Accordingly, the three symmetrized second-order elastic constants can be written as

$$\left.\begin{aligned} c_A &= c_A(0) + C'_{111}\eta_A \\ c_E &= c_E(0) + C'_{122}\eta_A \\ c_T &= c_T(0) + C'_{144}\eta_A \end{aligned}\right\} \tag{10–72}$$

where sub T actually means irrep T_2. (See Eqs. (9–38) to (9–40) for the meaning of the symmetrized constants in terms of the two-index engineering constants.) Thus, measurements of elastic constants as a function of hydrostatic deformation yield the three TOECs: C'_{111}, C'_{122}, and C'_{144}. From Eqs. (60)–(65) and (69) we can show that

$$C'_{111} = (1/\sqrt{3})(C^{e}_{111} + 6C^{e}_{112} + 2C^{e}_{123}) \tag{10–73}$$

$$C'_{122} = (1/\sqrt{3})(C^{e}_{111} - C^{e}_{123}) \tag{10–74}$$

$$C'_{144} = (2/\sqrt{3})(C^{e}_{144} + 2C^{e}_{155}) \tag{10–75}$$

Accordingly, such pressure measurements allow us to obtain the three combinations of TOECs: $C^{e}_{111} + 2C^{e}_{112}$, $C^{e}_{144} + 2C^{e}_{155}$ and $C^{e}_{111} - C^{e}_{123}$. Clearly, for the remaining three TOECs, applied static strains other than η_A are required (i.e. those involving shear strains, namely, η_E and η_T).

A second method for obtaining TOECs, introduced more recently, does not require the imposition of static strains. This method is based on ultrasonic harmonic generation. The principle is that an ultrasonic wave of finite amplitude becomes distorted as it progresses and higher harmonics are generated. The measurement of the amplitude of the harmonics offers a unique method to determine TOECs. The details of this method are beyond the scope of this book, but they are fully covered by Breazeale and Philip (1984). This method is especially useful for measurement of TOECs as a function of temperature, particularly down to cryogenic temperatures. Most of the studies using it were carried out on cubic crystals: simple metals (e.g. Cu), semiconductors (Si and Ge), alkali halides and some perovskites, representative values of which are given in Table 10–3. The best results for the six TOECs have been obtained by

Table 10–3. *TOECs of some cubic materials at room temperature (in units of* 10^{12} dyn/cm^2) *(from Breazeale and Philip, 1984)*

Material	C^e_{111}	C^e_{112}	C^e_{123}	C^e_{144}	C^e_{155}	C^e_{456}
Cu	−14.3	−8.8	−1.8	−0.6	−7.4	+0.66
Ge	−7.4	−3.9	−0.6	+0.1	−2.6	−1.1
Si	−8.3	−5.3	−0.02	−0.95	−3.0	−0.07
NaCl	−9.5	−0.8	+0.2	−0.8	+0.9	+0.2

combining results of harmonic-generation measurements with those from hydrostatic pressure dependence of ultrasonic wave velocities.

Thus far, there has only been a small amount of work on the upper uniaxial crystals which, of course, have more TOEC constants (10 for hexagonal, 12 for tetragonal and 14 for trigonal).

10–3 Other cases of $T(6)$ tensors

While no other cases of $T(6)$ tensors are as widely studied as the TOEC, the possibilities for matter tensors of this rank are, in fact, quite numerous. In this section, we will briefly mention some of these possibilities and indicate how they should be attacked.

First, let us consider the relation between a $T_S(2)$ and a $T_S(4)$, as in Section 10–1, except that now the resulting matter tensor K_{ikl}, expressed in three-index form, is only symmetric with respect to exchange of k and l, but not with respect to interchange of all three indices. In this case, the maximum number of coefficients is 6×21, or 126. An example of such a relationship comes from the dependence of elastic constants c_{ij} (a $T_S(4)$ quantity), on direction of magnetization in a magnetic crystal through the second-order products of direction cosines (a $T_S(2)$ quantity). Such a $T(6)$ tensor is easily handled by simply following the methods of Section 10–1 *without* invoking the additional condition of interchange of all three indices in K'_{ikl}.

Another example (Keyes, 1960) is the effect of stress (a $T_S(2)$ quantity) on the magnetoresistivity tensor ξ (a $T(4)$ quantity) to produce a $T(6)$ tensor. Note that ξ_{ijkl} is symmetric only with respect to $i \leftrightarrow j$ and $k \leftrightarrow l$ but not for $ij \leftrightarrow kl$ (see Section 2–4). This tensor then has 36 components, so that the $T(6)$ matter tensor we are seeking has $6 \times 36 = 216$ components. In handling such a tensor, we can split the $T(4)$ tensor into the sum of symmetric and antisymmetric products of $T_S(2)$ quantities (with 21 and 15 components respectively), in a manner analogous to what we did with

Table 10–1. *The Symmetry-Coordinate Transformation tables for quantities α_i and δ'_j. (The δ'_j are the symmetric products of the symmetry coordinates $\bar{\alpha}_i$ as defined by Eqs. (3) and (4); for uniaxial and cubic groups the $\bar{\alpha}_i$ are taken in sequence from the first column of this table. For orthorhombic and lower symmetries, $\bar{\alpha}_i = \alpha_i$.)*

Monoclinic

C_s		
A'	α_1; α_2; α_3; α_6	δ'_1 through δ'_8; δ'_{11}; δ'_{12}; δ'_{15}; δ'_{18}; δ'_{19}
A''	α_4; α_5	δ'_9; δ'_{10}; δ'_{13}; δ'_{14}; δ'_{16}; δ'_{17}; δ'_{20}; δ'_{21}

Orthorhombic

C_{2v}		
A_1	α_1; α_2; α_3	δ'_1 through δ'_8; δ_{12}
A_2	α_6	δ'_{11}; δ'_{15}; δ'_{18}; δ'_{19}
B_1	α_5	δ'_{10}; δ'_{14}; δ'_{17}; δ'_{20}
B_2	α_4	δ'_9; δ'_{13}; δ'_{16}; δ'_{21}

Trigonal (upper)

C_{3v}		
A_1	$\alpha_1 + \alpha_2$; α_3	δ'_1; δ'_2; δ'_7; $\delta'_3 + \delta'_4$; $\delta'_5 + \delta'_6$; $\delta'_{17} + \delta'_{20}$
A_2	–	$\delta'_{19} - \delta'_{18}$
E	(α_4, α_5); $(\alpha_1 - \alpha_2, \alpha_6)$	$(\delta'_8; \delta'_9)$; $(\delta'_{10}, \delta'_{11})$; $(\delta'_{12}, \delta'_{13})$; $(\delta'_{14}, \delta'_{15})$; $(\delta'_4 - \delta'_3, \delta'_{16})$; $(\delta'_5 - \delta'_6, -\delta'_{21})$; $(\delta'_{20} - \delta'_{17}, \delta'_{18} + \delta'_{19})$

Trigonal (lower)

C_3		
A	$\alpha_1 + \alpha_2$; α_3	same as C_{3v}, irreps A_1 and A_2
$\tilde{E}$	$\alpha_4 \pm i\alpha_5$; $\alpha_1 - \alpha_2 \pm i\sqrt{2}\alpha_6$	same as C_{3v}, irrep E: first $\pm$ i (second)

Tetragonal (upper)

C_{4v}		
A_1	$\alpha_1 + \alpha_2$; α_3	δ'_1; δ'_2; δ'_3; δ'_4; $\delta'_5 + \delta'_6$; δ'_7
A_2	–	δ'_{16}
B_1	$\alpha_1 - \alpha_2$	δ'_8; δ'_{12}; $\delta'_5 - \delta'_6$
B_2	α_6	δ'_9; δ'_{13}; δ'_{21}
E	(α_4, α_5)	$(\delta'_{10}, \delta'_{11})$; $(\delta'_{14}, \delta'_{15})$; $(\delta'_{20}, \delta'_{19})$; $(\delta'_{17}, -\delta'_{18})$

Table 10–1. *(cont.)*

Tetragonal (lower)

C_4		
A	$\alpha_1 + \alpha_2$; α_3	same as C_{4v}, irreps A_1 and A_2
B	$\alpha_1 - \alpha_2$; α_6	same as C_{4v}, irreps B_1 and B_2
$\tilde{E}$	$\alpha_4 \pm i\alpha_5$	same as C_{4v}, irrep E_1: first $\pm$ i (second)

Hexagonal (upper)

C_{6v}		
A_1	$\alpha_1 + \alpha_2$; α_3	δ'_1; δ'_2; $\delta'_3 + \delta'_4$; $\delta'_5 + \delta'_6$; δ'_7
A_2	–	–
B_1	–	$\delta'_{17} + \delta'_{20}$
B_2	–	$\delta'_{19} - \delta'_{18}$
E_1	(α_4, α_5)	(δ'_8, δ'_9); $(\delta'_{12}, \delta'_{13})$; $(\delta'_{20} - \delta'_{17}, \delta'_{18} + \delta'_{19})$
E_2	$(\alpha_1 - \alpha_2, \alpha_6)$	$(\delta'_{10}, \delta'_{11})$; $(\delta'_{14}, \delta'_{15})$; $(\delta'_4 - \delta'_3, \delta'_{16})$; $(\delta'_5 - \delta'_6, -\delta'_{21})$

Hexagonal (lower)

C_6		
A	$\alpha_1 + \alpha_2$; α_3	same as C_{6v}, irrep A_1
B	–	same as C_{6v}, irreps B_1 and B_2
$\tilde{E}_1$	$\alpha_4 \pm i\alpha_5$	same as C_{6v}, irrep E_1: first $\pm$ i (second)
$\tilde{E}_2$	$\alpha_1 - \alpha_2 \mp i\sqrt{2}\alpha_6$	same as C_{6v}, irrep E_2: first $\mp$ i (second)

Cubic (upper)

O		
A_1	$\alpha_1 + \alpha_2 + \alpha_3$	δ'_1; $\delta'_2 + \delta'_3$; $\delta'_4 + \delta'_5 + \delta'_6$
A_2	–	–
E	$(2\alpha_1 - \alpha_2 - \alpha_3, \alpha_2 - \alpha_3)$	(δ'_7, δ'_8); $(\delta'_2 - \delta'_3, -\delta'_{12})$; $(2\delta'_4 - \delta'_5 - \delta'_6, \delta'_5 - \delta'_6)$
T_1	–	$(\delta'_{16}, -\sqrt{3}\delta'_{14} - \delta'_{17}, \sqrt{3}\delta'_{15} - \delta'_{18})$
T_2	$(\alpha_4, \alpha_5, \alpha_6)$	$(\delta'_9, \delta'_{10}, \delta'_{11})$; $(-\delta'_{13}, \delta'_{14} - \sqrt{3}\delta'_{17}, \delta'_{15} + \sqrt{3}\delta'_{18})$; $(\delta'_{21}, \delta'_{20}, \delta'_{19})$

Cubic (lower)

T		
A	$\alpha_1 + \alpha_2 + \alpha_3$	same as O, irrep A_1
$\tilde{E}$	$(2\alpha_1 - \alpha_2 - \alpha_3 \mp i(\alpha_2 - \alpha_3))$	same as O, irrep E: first $\mp$ i (second)
T	$(\alpha_4, \alpha_5, \alpha_6)$	same as O, irreps T_1 and T_2

the products of vectors in Section 3–4. In this way, we can utilize the material that we have developed earlier in this chapter in handling the symmetric part. The antisymmetric part would then require separate treatment.

Still another example is second-order magnetostriction, the dependence of stress on fourth powers of the direction of magnetization of a magnetic material (Birss, 1964). In this case we have the relationship of a $T_S(2)$ tensor to a $T_S(4)$ that is *fully symmetric*, since it corresponds to all products of vector components to the fourth power. Thus, in X_{ijkl} not only do we have symmetry with respect to $i \leftrightarrow j$, $k \leftrightarrow l$, and $ij \leftrightarrow kl$, as we had in Section 10–1, but now we also add $i \leftrightarrow k$. This $T_S(4)$ then has only 15 components and the maximum number for the corresponding matter tensor **K** is $6 \times 15 = 90$ components. This problem can be handled by introducing appropriate equalities among the quantities δ_k of Eq. (3), namely (see Problem 10–6),

$$\left.\begin{array}{lll} \delta_6 = \delta_7; & \delta_5 = \delta_8; & \delta_4 = \delta_{12}; \\ \delta_9 = \delta_{21}; & \delta_{14} = \delta_{20}; & \delta_{18} = \delta_{19}. \end{array}\right\} \qquad (10\text{–}76)$$

Another alternative for obtaining the components of a fully symmetric $T_S(4)$ is to develop S-C-T tables for quadruple products of a vector with itself, analogous to the α's (double products) and β's (triple products) that we have already presented in Appendix E.

From these considerations, the reader may see that, while obtaining the effects of symmetry on high-order tensors can become a formidable task, the methods that we have developed herein can provide solutions to such problems.

Problems

10–1. Obtain the symmetry coordinates of δ_i' and their irreps for the upper cubics (group O) upper tetragonals (C_{4v}) and upper hexagonals (C_{6v}), including similarity of orientation to the corresponding $\bar{\alpha}_i$ in the case of E or T irreps. Compare your results with Table 10–1.

10–2. Do the same for the lower hexagonal case (C_6). (Here the question of similarity of orientation does not arise.)

10–3. Following the same procedure as in Section 10–1, obtain the three-index **K** tensor for the lower cubic and for the upper and lower tetragonal groups. Compare with Table 10–2.

10–4. Show that the method for combining the symmetry of the upper cubic and upper hexagonal groups used to obtain Eqs. (66)–(68) for the isotropic case may also be used for the tensor that relates a $T_S(2)$ to a $T_S(2)$, to give Eq. (9–6).

10–5. For an upper hexagonal crystal, show which TOEC coefficients may be obtained from measurements of the effect of hydrostatic pressure on the elastic constants c_{ij}. (Hint: first show that hydrostatic pressure gives rise to two different Lagrangian strains belonging to irrep A_1.)

10–6. Verify Eq. (76) for the equalities among the δ_i resulting from full symmetry of the $T_S(4)$ tensor.

Appendix A

Review of tensors

A–1 Linear orthogonal transformations

Consider a rectangular cartesian coordinate system X (axes OX_1, OX_2, OX_3) in three dimensions. A given point P can then be described by its coordinates x_1, x_2, x_3 in this system. Now consider a change to another rectangular coordinate system X' with the same origin. The same point P is described in the new system by coordinates x'_1, x'_2, x'_3, such that

$$\left.\begin{aligned} x'_1 &= a_{11}x_1 + a_{12}x_2 + a_{13}x_3 \\ x'_2 &= a_{21}x_1 + a_{22}x_2 + a_{23}x_3 \\ x'_3 &= a_{31}x_1 + a_{32}x_2 + a_{33}x_3 \end{aligned}\right\} \qquad \text{(A–1)}$$

or, more concisely, as

$$x'_i = a_{ij}x_j \quad (i, j, = 1, 2, 3) \qquad \text{(A–2)}$$

using the Einstein summation convention (see page 2). It is easy to see that the quantities a_{ij} are direction cosines between the first (unprimed) and second (primed) sets of axes. For this reason, the inverse transformation may be written

$$x_i = a_{ji}x'_j \qquad \text{(A–3)}$$

that is, by changing rows into columns in Eqs. (1). The transformation may also be written in matrix form:

$$\mathbf{r}' = \mathbf{A}\mathbf{r} \qquad \text{(A–4)}$$

where $\mathbf{r}'$ and $\mathbf{r}$ are in column form:

$$\mathbf{r}' = \begin{pmatrix} x'_1 \\ x'_2 \\ x'_3 \end{pmatrix}, \mathbf{r} = \begin{pmatrix} x_1 \\ x_2 \\ x_3 \end{pmatrix} \qquad \text{(A–5)}$$

and $\mathbf{A}$ is the matrix of a_{ij} coefficients. The inverse transformation is

$$\mathbf{r} = \mathbf{A}^{-1}\mathbf{r}' = \mathbf{A}^{\dagger}\mathbf{r}' \qquad \text{(A–6)}$$

where $\mathbf{A}^{\dagger}$ denotes the transpose of the matrix $\mathbf{A}$.* This result, that the inverse matrix $\mathbf{A}^{-1}$ equals the transposed $\mathbf{A}^{\dagger}$, characterizes a linear orthogonal transformation.

* More generally, when $\mathbf{A}$ consists of terms that are complex quantities, the symbol $\mathbf{A}^{\dagger}$ will be used to represent the *transposed complex conjugate* of $\mathbf{A}$. When $\mathbf{A}^{-1} = \mathbf{A}^{\dagger}$, matrix $\mathbf{A}$ is then said to be unitary. In the present case, however, $\mathbf{A}$ is made up of real quantities and so the † symbol only means transposed.

From the fact that the a_{ij} are sets of direction cosines, it follows that the sum of the squares of the coefficients in each of Eqs. (1) is equal to one and that the 'dot product' of coefficients of two different rows equals zero. In compact notation, this may be written:

$$a_{ik}a_{jk} = \delta_{ij} \tag{A-7}$$

where δ_{ij} is the Kronecker delta ($\delta_{ij} = 1$ if $i = j$, $\delta_{ij} = 0$ if $i \neq j$). Similarly, from the inverse transformation, one obtains

$$a_{ki}a_{kj} = \delta_{ij} \tag{A-8}$$

If two linear orthogonal transformations are applied successively, the first taking x_j into x'_i in accordance with Eq. (1) and the second, taking the x'_j into x''_i according to

$$x''_i = b_{ij}x'_j \tag{A-9}$$

we wish to determine the direct transformation from the x_j into the x''_i. If this is written as

$$x''_i = c_{ij}x_j \tag{A-10}$$

then it is readily shown that

$$c_{ij} = b_{ik}a_{kj} \tag{A-11}$$

In matrix form, the total transformation matrix **C** is

$$\mathbf{C} = \mathbf{BA} \tag{A-12}$$

that is, it is the matrix product of the two successive transformation matrices. Further, if **A** and **B** are orthogonal matrices, so is **C**.

Next we wish to consider the determinant of the transformation **A**. Since **A** followed by $\mathbf{A}^{-1}$ gives the identity matrix

$$\mathbf{I} = \begin{pmatrix} 1 & 0 & 0 \\ 0 & 1 & 0 \\ 0 & 0 & 1 \end{pmatrix}$$

and since $\mathbf{A}^{-1} = \mathbf{A}^{\dagger}$, we have

$$\mathbf{AA}^{\dagger} = \mathbf{I} \tag{A-13}$$

Taking determinants of Eq. (13), noting that the determinant of a product is the product of the determinants, and also that $\det \mathbf{A} = \det \mathbf{A}^{\dagger}$ (from the definition of determinants), we obtain

$$(\det \mathbf{A})^2 = 1 \tag{A-14}$$

and therefore, $\det \mathbf{A} = \pm 1$. This result is an important characteristic of an orthogonal matrix **A**. In fact, there are two types of orthogonal transformation. Those for which $\det \mathbf{A} = +1$ preserve the sense or 'handedness' of the axes, for example take a right-handed set into another right-handed set. Such a transformation is a pure rotation of the axes. The other type, for which $\det \mathbf{A} = -1$, reverses the handedness of the axes and includes reflections, inversions and rotation–reflections (rotation followed by, or preceded by, a reflection or inversion).

A–2 Defining tensors

A quantity p that is invariant under an orthogonal transformation of coordinates is said to be a *scalar* or a *tensor of rank* zero, $T(0)$. Thus, for a scalar,

$$p' = p \tag{A-15}$$

where the prime denotes the value in the new coordinate system X' obtained by the transformation of Eq. (1).

A set of three quantities p_i ($i = 1, 2, 3$) that transform in the same manner as the coordinates is said to be a (polar) *vector* or *tensor of rank 1*, $T(1)$. Thus,

$$p_i' = a_{ij}p_j \quad (i, j = 1, 2, 3) \tag{A-16}$$

where the unprimed and primed quantities again denote the values in the original and in the new coordinate system respectively. The quantities p_1, p_2 and p_3 are called the three *components* of the vector. These may again be represented as a column matrix **p** in the manner of Eq. (5).

For later purposes, it is also worth pointing out how products of vector components transform. Thus, given two vectors **p** and **q** and taking the nine products p_iq_j, it follows from Eq. (16) that these products transform as

$$p_i'q_j' = (a_{ik}a_{jl})p_kq_l \tag{A-17}$$

In a similar way triple products of components of vectors **p**, **q** and **r** transform as

$$p_i'q_j'r_k' = (a_{il}a_{jm}a_{kn})p_lq_mr_n \tag{A-18}$$

and so on for higher products.

A set of quantities p_{ij} with nine components is said to be a *tensor of rank 2*, denoted $T(2)$, if it transforms under an orthogonal transformation in the same manner as the set of products of two vectors in Eq. (17), namely,

$$p_{ij}' = a_{ik}a_{jl}p_{kl} \tag{A-19}$$

We are now ready to extend the definition to a tensor of rank n, denoted by $T(n)$ as a set of 3^n components $p_{ijk\ldots}$ with n indices, transforming according to

$$p_{ijk\ldots}' = a_{ir}a_{js}a_{kt}\ldots p_{rst\ldots} \tag{A-20}$$

that is, in the same manner as the set of all products of n vector components with each other.

A–3 Algebra of tensors

From the transformation property of tensors, it is straightforward to show the following theorems:

1. Any linear combination of tensors of rank n is a tensor of rank n. (This includes addition and subtraction of two $T(n)$ tensors and multiplication of a $T(n)$ by a constant.)
2. If **p** or $p_{ij\ldots}$ is a $T(m)$ and **q** or $q_{kl\ldots}$ is a $T(n)$, then the set of products of all p's and q's forms a tensor of rank $m + n$. Such a product of **p** and **q** is called an *outer product*.
3. Products can also be carried out in which one index is repeated, so that there is a contraction of the resulting tensor. For example, if this is done for the product of **p** and **q** of item (2), the resulting tensor will have rank $m + n - 2$. In general, the rank is reduced by 2 for every index summed over. This operation is called an *inner product*.

A special case is the scalar or 'dot' product of two vectors, p_iq_i, which is, of course, a scalar. Since a $T(n)$ is the equivalent of a product of n vectors, it is clear why in an inner product each contraction that eliminates an index produces a scalar and reduces the rank by 2.

Conversely, the above considerations allow one to identify the rank of a tensor relating two other tensors. For example, consider the relation

$$p_{ij} = q_{ijkl} r_{kl} \tag{A–21}$$

between two arbitrary $T(2)$ quantities p_{ij} and r_{kl}. Here we have contraction of two indices. It can be shown that q_{ijkl} must be a $T(4)$ tensor, since only if it is, can p_{ij} have the assumed $T(2)$ transformation properties.

A generalization of the above allows us to identify readily the tensor character of a set of coefficients relating two tensors of known rank.

A–4 Symmetry of tensors

If an interchange of the same two indices of each tensor component leaves its value unchanged, the tensor is said to be *symmetric* with respect to these indices. Thus, a symmetric tensor of rank two has

$$p_{ij} = p_{ji} \quad (\text{all } i, j) \tag{A–22}$$

and, therefore, has only six independent components.

If the interchange of the same two indices i and j always results in a reversal of sign of all components, the tensor is said to be *antisymmetric* in the indices i and j. In the special case of a $T(2)$, antisymmetry means

$$\left.\begin{array}{ll} p_{ij} = -p_{ji} & (i \neq j) \\ p_{ii} = 0 & (i = 1, 2, 3) \end{array}\right\} \tag{A–23}$$

so that there are only three independent components. From the transformation properties of tensors, the following theorem is readily demonstrated:

The symmetry property of a tensor is preserved under an orthogonal transformation.

A–5 Representation quadric of a symmetric $T(2)$

Symmetric tensors of rank 2 occur very frequently in crystal physics: on the one hand, for forces and responses, notably stress and strain, respectively, and, on the other hand, for crystal properties such as dielectric permeability, electrical conductivity, etc.

Consider such a tensor p_{ij} and let us write the quadratic form

$$p_{ij} x_i x_j = 1 \tag{A–24}$$

based on the p_{ij} coefficients. This equation represents a quadric surface (ellipsoid or hyperboloid) in the three-dimensional space of the coordinate system X. Equation (24) may be transformed to a new coordinate system X' by carrying out an orthogonal transformation:

$$p'_{ij} x'_i x'_j = 1 \tag{A–25}$$

where the p'_{ij} are given by Eq. (19) and the x'_i and x'_j by Eq. (2). An important property of such a symmetric matrix $\mathbf{p}$ is that there always exists an orthogonal transformation that will take it into diagonal form

$$\mathbf{p}' = \begin{pmatrix} p_1 & 0 & 0 \\ 0 & p_2 & 0 \\ 0 & 0 & p_3 \end{pmatrix} \tag{A–26}$$

that is, in which all off-diagonal elements are zero. The three diagonal elements are known as the 'principal values'.

The corresponding quadric in this set of axes takes the form

$$p_1x_1'^2 + p_2x_2'^2 + p_3x_3'^2 = 1 \tag{A-27}$$

and the axes X' are called the 'principal axes'. It is noteworthy that, if all three principal values are positive, the quadric surface is an ellipsoid, which in standard form can be written

$$\frac{x_1'^2}{a^2} + \frac{x_2'^2}{b^2} + \frac{x_3'^2}{c^2} = 1 \tag{A-28}$$

where a, b and c are the semiaxes. Thus, $a = 1/\sqrt{p_1}$, and so on. If one or two of the principal values are negative, hyperboloids are obtained.

In matrix terminology, the transformation of matrix $\mathbf{p}$ to principal values or diagonal form $\mathbf{p}'$ by an orthogonal transformation $\mathbf{A}$ may be written

$$\mathbf{ApA}^\dagger = \mathbf{p}' \tag{A-29}$$

Such a matrix transformation, which is equivalent to Eq. (19), is called a 'similarity transformation' of $\mathbf{p}$.

A–6 Axial tensors

An *axial vector*, denoted by $T(1)^{\text{ax}}$ is a set of three components p_i which, under the orthogonal transformation of Eq. (2), transform to

$$p_i' = \pm a_{ij}p_j \tag{A-30}$$

with the $\pm$ signs according as $\det \mathbf{A} = \pm 1$. In other words, the axial vector components transform like a true vector (or polar vector) under a transformation that preserves the handedness of the axes (i.e. a rotation) but, unlike a vector, reverse their sign under a transformation that changes the handedness of the axes (e.g. a rotation–reflection).

In a similar way, an axial tensor of rank n, denoted by $T(n)^{\text{ax}}$, has the transformation property

$$p'_{ijk\ldots} = \pm a_{ir}a_{js}a_{kt}\ldots p_{rst\ldots} \tag{A-31}$$

where the $\pm$ signs are again as $\det \mathbf{A} = \pm 1$. Axial tensors are also called 'pseudotensors'.

Axial vectors can also be viewed as the 'cross product' of two polar vectors. Thus, consider vectors $\mathbf{h}$ and $\mathbf{k}$ and their cross product

$$\mathbf{p} = \mathbf{h} \times \mathbf{k} \tag{A-32}$$

which, in component form, is

$$\left.\begin{aligned} p_1 &= h_2k_3 - h_3k_2 \\ p_2 &= h_3k_1 - h_1k_3 \\ p_3 &= h_1k_2 - h_2k_1 \end{aligned}\right\} \tag{A-33}$$

Under an orthogonal transformation, Eq. (2), the components of $\mathbf{p}$ transform as

$$\begin{aligned} p_1' = (a_{22}a_{33} - a_{23}a_{32})p_1 + (a_{23}a_{31} - a_{21}a_{33})p_2 \\ + (a_{21}a_{32} - a_{22}a_{31})p_3 \end{aligned} \tag{A-34}$$

and so on for p'_2 and p'_3. Clearly, **p** is not a vector. In fact, if we write the **p** components in two-index notation:

$$p_{12} \to -p_3;\ p_{23} \to -p_1;\ p_{31} \to -p_2 \qquad \text{(A–35)}$$

so that

$$p_{ij} = -h_i k_j + h_j k_i \qquad \text{(A–36)}$$

it becomes clear that the p_{ij} are components of a second-rank antisymmetric tensor which takes the form

$$\mathbf{p} = \begin{pmatrix} 0 & -p_3 & p_2 \\ p_3 & 0 & -p_1 \\ -p_2 & p_1 & 0 \end{pmatrix} \qquad \text{(A–37)}$$

Nevertheless, since the following relations hold for the transformation coefficients a_{ij}:

$$\left.\begin{aligned} a_{11} &= \pm(a_{22}a_{33} - a_{23}a_{32}) \\ a_{12} &= \pm(a_{23}a_{31} - a_{21}a_{33}) \\ a_{13} &= \pm(a_{21}a_{32} - a_{22}a_{31}) \end{aligned}\right\} \qquad \text{(A–38)}$$

(where the $\pm$ correspond to $\det \mathbf{A} = \pm 1$), comparison with Eq. (34) shows that the single-index quantities p_i also transform as an axial vector. (Note that each quantity in parentheses on the right-hand side of Eqs. (38) is the minor of the quantity on the left in $\det \mathbf{A}$.)

Just as a polar vector is symbolized by an arrow indicating direction, it is convenient to symbolize an axial vector by a line with a sense of rotation, as in Fig. A–1. These diagrams indicate the transformation properties, namely, under a pure rotation both figures transform in the same way, but under reflection they behave oppositely. For example, reflection through a plane perpendicular to the line of the vector reverses the direction of a polar vector but leaves the axial vector unchanged, while reflection through a plane containing the vector has the opposite effect on both.

Returning to considerations of axial tensors of arbitrary rank, as defined by the transformation Eq. (31), it is easy to see that, because of the $\pm$ sign, the outer product of an *even* number of axial tensors is a true vector of appropriate rank. On the other hand, the outer product of an *odd* number of axial tensors remains

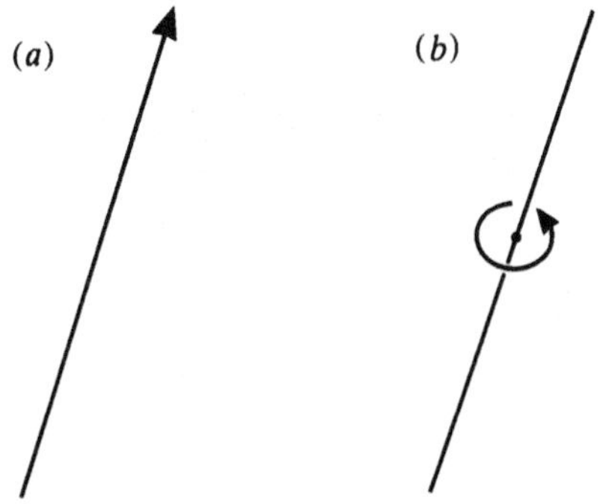

Fig. A–1. Diagrammatic representation of (*a*) a polar vector, and (*b*) an axial vector.

an axial tensor. As for true tensors, the rank is obtained as the sum of the ranks of the tensors being multiplied.

One limiting case of Eq. (31) deserves special mention, namely, the case of $n = 0$. Thus, a quantity p is called a *pseudoscalar*, or, more formally, an axial tensor of rank zero, $T(0)^{ax}$, if it has the property

$$p' = \pm p \tag{A-39}$$

that is, maintaining both its value and sign under rotations, but reversing sign under reflections or rotation-reflections. It is sometimes convenient to think of an axial tensor of order n as having the transformation properties of the product of n vectors and a pseudoscalar.

Appendix B

Stress, strain and elasticity

B–1 Stress

The state of stress at point $P(x_1, x_2, x_3)$ in an elastic medium will be shown to be a symmetric tensor of second rank. Let us first set up an infinitesimal parallelopiped at P with edges δx_1, δx_2, δx_3, parallel to the three coordinate axes, respectively, as shown in Fig. B–1. There will be a stress (force/area) acting on each of the six faces, each of which may be resolved into components parallel to the three axes. Furthermore, for static equilibrium, the forces on opposite faces must be equal and opposite. Thus, we need only give the nine components, σ_{ij}, as shown in Fig. B–1. The terminology is such that σ_{ij}, is the force/area on a plane perpendicular to axis i acting along direction j. A further requirement for static equilibrium is that there must be no net torque about each of the axes. Accordingly,

$$\sigma_{ij'} = \sigma_{ji} \tag{B–1}$$

Thus σ_{ij} is symmetric and has only six independent components.

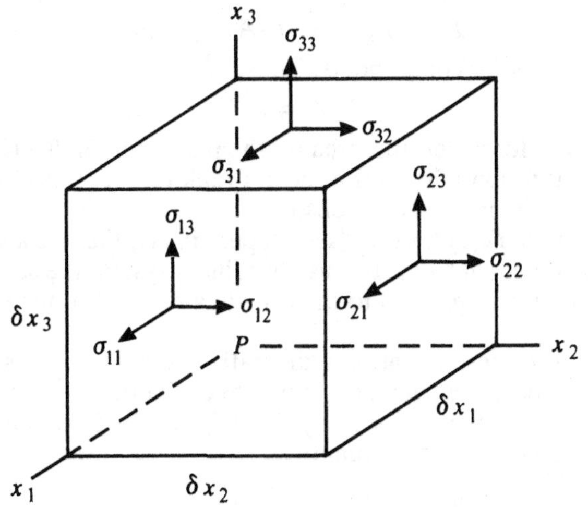

Fig. B–1. Representation of a state of stress in a medium, showing the stresses acting on three of the faces of an infinitesimal rectangular parallelopiped at point P.

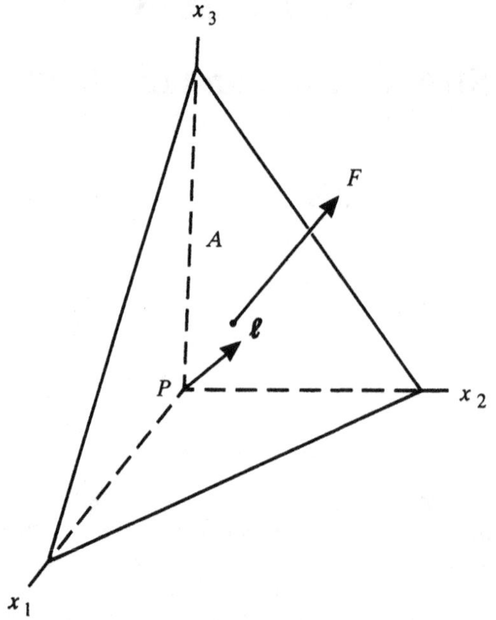

Fig. B–2. Representation of the state of stress, showing the force acting on a small area A perpendicular to direction $\boldsymbol{l}$ emerging from point P.

An important property of the system of stresses σ_{ij}, is that the force/area about any direction designated by unit vector $\boldsymbol{l}$, emerging from the reference point P is readily obtainable in terms of the σ_{ij}. Let A be a small area perpendicular to direction $\boldsymbol{l}$, as shown in Fig. B–2 and $\mathbf{F}$ be the vector force acting on that area. Taking A_i, A_2, and A_3 as the three projections of A perpendicular to the coordinate axes, with $A_i = l_i A$, then, clearly, component F_1 is given by

$$F_1 = \sigma_{11}A_1 + \sigma_{12}A_2 + \sigma_{13}A_3 \qquad \text{(B–2)}$$

and similarly for the other components. Thus,

$$F_i/A = \sigma_{ij}l_j \qquad \text{(B–3)}$$

gives the component force per unit area on the plane perpendicular to $\boldsymbol{l}$. Since $\mathbf{F}$ and $\boldsymbol{l}$ are vectors, it follows that σ is a second-rank tensor. It is also symmetric in view of Eq. (1), that is, it is a $T_S(2)$ tensor.

As in the case of all $T_S(2)$ tensors (see Appendix A), there is a set of axes for which $\boldsymbol{\sigma}$ takes the diagonal form. In this coordinate system the stresses, called the 'principal stresses', are all parallel to the coordinate axes and the shear stresses are all zero.

A special case of interest is that of hydrostatic stress. Here the stress tensor is diagonal with all three principal values equal to each other. It can therefore be written as a scalar quantity multiplied by the unit matrix. Such a hydrostatic stress can, in fact, be treated as a scalar quantity.

B–2 Strain

The strain tensor is a set of values that describes the local distortion of the medium in the vicinity of a given point. To define strain, we consider the change in relative position of two adjacent points P and Q, initially at (x_1, x_2, x_3) and $(x_1 + dx_1, x_2 + dx_2, x_3 + dx_3)$ respectively. Under a state of strain, P is displaced to P′ whose coordinates are $x_i + u_i$ $(i = 1, 2, 3)$ while Q goes to Q', for which the x_1 coordinate is

$$(x_1 + dx_1) + u_1 + \left(\frac{\partial u_1}{\partial x_1}\right)dx_1 + \left(\frac{\partial u_1}{\partial x_2}\right)dx_2 + \left(\frac{\partial u_1}{\partial x_3}\right)dx_3$$

and similarly for the other two components. We may write these coordinates as $x_i + dx_i + u_i + du_i$, where

$$du_1 = \frac{\partial u_i}{\partial x_j}dx_j \tag{B-4}$$

using the summation convention. The set of nine components $\partial u_i/\partial x_j$ are, in general, functions of position, that is, of x_1, x_2, x_3. The set of quantities du_i, representing the additional displacement of Q relative to P, constitute a vector quantity, as does the set dx_i. Thus $\partial u_i/\partial x_j$ constitutes a tensor of second rank, $T(2)$. As in the case of any $T(2)$, this tensor can be written as the sum of a symmetric and an antisymmetric second-rank tensor by defining

$$\varepsilon_{ij} \equiv \tfrac{1}{2}\left(\frac{\partial u_i}{\partial x_j} + \frac{\partial u_j}{\partial x_i}\right) \tag{B-5}$$

for the symmetric tensor, and

$$\omega_{ij} \equiv \tfrac{1}{2}\left(\frac{\partial u_i}{\partial x_j} - \frac{\partial u_j}{\partial x_i}\right) \tag{B-6}$$

for the antisymmetric. The antisymmetric ω_{ij} can readily be shown to represent a rigid-body rotation of the elemental volume surrounding point P. It is, therefore, not a deformation and is of no interest for elasticity. The $T_S(2)$ tensor ε_{ij} is the strain tensor that we seek. From the definition, Eqs (4) and (5), it is clear that the diagonal components ε_{ii} represent the change in length per unit length along axis i, that is, an axial tensile or compressive strain, according as it is > 0 or < 0. The quantity ε_{ij} $(i \neq j)$, on the other hand, represents half the change in angle

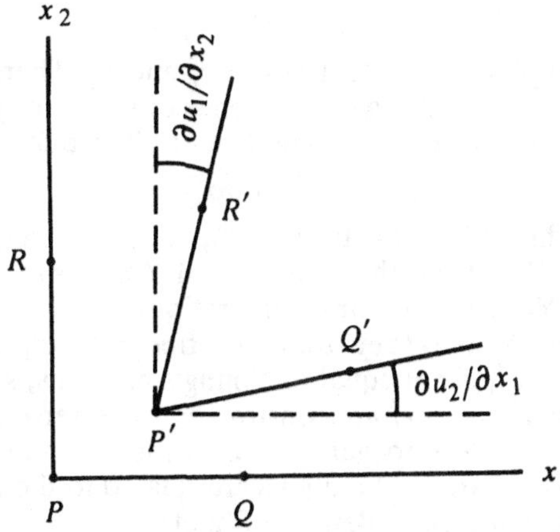

Fig. B–3. A two-dimensional diagram showing the onset of a shear strain ε_{12}. Here P, Q, R and P', Q', R' are, respectively, the positions of three points before and after the deformation. The quantity ε_{12} is one-half the change in angle between OX_1 and OX_2.

between axes OX_i and OX_j through the point P. It is called a shear strain and matches the directions of the stresses σ_{ij} shown in Fig. B–1. Figure B–3 shows the shear strain ε_{12} in a two-dimensional diagram. Here we use two points adjacent to P: point Q such that $PQ = dx_1$ and R such that $PR = dx_2$. The displacement of Q to Q' gives $\partial u_2/\partial x_1$ and that of R to R' gives $\partial u_1/\partial x_2$.

It is straightforward to show that for a direction $\boldsymbol{l}$ passing through P,

$$\varepsilon = \varepsilon_{ij} l_i l_j \tag{B–7}$$

is the change of length per unit length of a line element passing through P. The quantity ε, representing this tensile strain, is a scalar since ε_{ij} is a second-rank tensor and l_i is a vector.

B–3 Elasticity

Hooke's law, in its generalized form, states that each component of the strain tensor is a linear combination of all six stress tensor components, and vice versa. Thus,

$$\varepsilon_{ij} = s_{ijkl}\sigma_{kl} \tag{B–8}$$

where the set of quantities s_{ijkl} comprise the components of the *elastic compliance tensor*, a tensor of fourth rank (since $\boldsymbol{\varepsilon}$ and $\boldsymbol{\sigma}$ are both of rank 2). Inverting Eq. (8) gives

$$\sigma_{ij} = c_{ijkl}\varepsilon_{kl} \tag{B–9}$$

in which the c_{ijkl} are components of the *elastic stiffness tensor* (also of rank 4). Clearly, the c_{ijkl} coefficients are related to the s_{ijkl} by determinantal equations, but these relations are not simple unless many of the coefficients are zero.

In Section 1–4, we show how the number of indices is reduced in stress–strain relations by defining the six engineering stresses and strains, σ_i^e and ε_i^e $(i, j = 1, \ldots, 6)$ so that Eqs. (8) and (9) become

$$\varepsilon_i^e = s_{ij}^e\sigma_j^e \tag{B–10}$$

and

$$\sigma_i^e = c_{ij}^e\varepsilon_j \tag{B–11}$$

with 'engineering' compliances and stiffnesses s_{ij}^e and c_{ij}^e. There are 21 independent s and c constants, as a consequence of intrinsic symmetry. Let us look at the meanings of some of the s_{ij}^e coefficients. The quantity s_{11}^e is given by

$$s_{11}^e = (\partial\varepsilon_1^e/\partial\sigma_1^e) \tag{B–12}$$

under conditions in which all other stresses, $\sigma_2^e, \ldots, \sigma_6^e$ are equal to 0. This quantity is clearly equal to $1/Y_1$, the reciprocal of Young's modulus in the direction OX_1, since Young's modulus is the ratio stress/strain in a given direction when all other stresses are zero (but strains are free to take place). The corresponding quantity c_{11}^e is not equal to Young's modulus, since it equals $(\partial\sigma_1^e/\partial\varepsilon_1^e)$ when *strains* $\varepsilon_2^e, \ldots, \varepsilon_6^e$ are constrained to be zero. Such conditions are more readily obtained in the propagation of acoustic waves in solids, where the velocities of the waves in various directions are related to the c_{ij}^e coefficients. This matter will be discussed in greater detail in Chapter 9.

Appendix C

Finite deformation

The definition of the strain components for infinitesimally small deformations was given in Appendix B. This definition is suitable for dealing with the problems of ordinary elasticity. When deformations are not small, and 'finite strains' are involved, greater care is needed in defining the strain components. Again, let the position of our reference point P before deformation be at coordinates x_i and, after deformation, P′ at x_i'. We define the 'Lagrangian strains' η_{ij} with respect to the original coordinates by

$$\eta_{ij} = \tfrac{1}{2}\left(\frac{\partial u_i}{\partial x_j} + \frac{\partial u_j}{\partial x_i} + \frac{\partial u_k}{\partial x_i}\cdot\frac{\partial u_k}{\partial x_j}\right) \qquad \text{(C–1)}$$

with summation on k (following the summation convention). Note that this definition is similar to Eq. (B–5) for ε_{ij}, except that it includes higher-order derivatives. These strain components are not amenable to a simple interpretation, as were the ε_{ij}, but, like the ε_{ij}, they form a $T_S(2)$ tensor. In the limit of infinitesimal deformation, the last term of Eq. (1) vanishes and we return to the quantities ε_{ij}. As in the case of the infinitesimal strains, it is possible to express the internal energy of deformation per unit volume, U, as a power series in these Lagrangian strains (see Eq. (1–37)). The quantity U may also be written in a manner analogous to Eq. (1–1) as

$$\mathrm{d}U = t_{ij}\eta_{ij} + T\,\mathrm{d}S \qquad \text{(C–2)}$$

where the quantities t_{ij}, the conjugate variables to the η_{ij}, are called the 'thermodynamic tensions'. Equation (2) shows that

$$t_{ij} = \partial U/\partial\eta_{ij} \qquad \text{(C–3)}$$

Again, these t_{ij} quantities differ from the stresses σ_{ij} defined in Appendix B, but they clearly become the σ_{ij} for very small deformations. As already mentioned, for ordinary elasticity the Lagrangian strains and thermodynamic tensions are not needed. For the purposes of this book, only in the study of the third-order elastic constants, Section 1–6–2 and Chapter 10, will it be necessary to introduce finite deformation.

Appendix D
The great orthogonality theorem

This theorem, which is proved in most books on group theory, may be stated as follows. Express the αth irrep of a group $\mathscr{G}$ by the set of matrices $[D^{(\alpha)}(R)]_{ij}$ where R is one of the operations of the group, and take the dimensionality of this irrep to be l_α. Then the components of the various irreps of group $\mathscr{G}$ obey the equation

$$\sum_R [D^{(\alpha)}(R)]_{ij}^* [D^{(\beta)}(R)]_{pq} = (h/l_\alpha)\delta_{\alpha\beta}\delta_{ip}\delta_{jq} \qquad \text{(D–1)}$$

where * denotes the complex conjugate, h is the order of the group and the δ's are all Kronecker δ's.

This rather complicated equation may be interpreted as follows. The fact that the order of the group is h means that each irrep has h matrices. Suppose we treat the corresponding elements of these matrices, for example $[D^{(\alpha)}(R)]_{ij}$ for given i and j, as we vary the group operation R, as a vector in h-dimensional space. There are l_α^2 such vectors for irrep α and $\sum_\alpha l_\alpha^2$ in total. But group theory also tells us that $\sum_\alpha l_\alpha^2 = h$ (see Eq. (3–10)), so that there are h such vectors in all. What Eq. (1) states is that all members of this collection of h vectors are orthogonal to each other (defined so that complex conjugates are used in the dot products, where necessary). Further, it tells us that the magnitude of each such vector (the dot product with itself) equals h/l_α.

Next we turn to the *characters* $\chi^{(\alpha)}(R)$ of the various irrep matrices. From the definition of the character, Eq. (3–8), it readily follows from Eq. (1) that the characters obey their own orthogonality relation

$$\sum_R \chi^{(\alpha)}(R)^* \chi^{(\beta)}(R) = h\delta_{\alpha\beta} \qquad \text{(D–2)}$$

Again, we may think of the characters of each irrep α as forming an h-dimensional vector. Equation (2) finds that these vectors are all orthogonal to each other and that the magnitude of each is equal to h, the order of the group.

The two lemmas

For the proof of the Great Orthogonality Theorem, two important lemmas are required. Since these will later be utilized in proving the Fundamental Theorem (Appendix F), they will be stated here.

1. 'Schur's Lemma'. Any matrix which commutes with all matrices of an irrep must be a constant matrix. Thus, if R is a group operation, $\mathbf{D}(R)$ is an irrep of the group, and

$$\mathbf{D}(R)\mathbf{M} = \mathbf{M}\mathbf{D}(R) \quad \text{for all } R$$

then $\mathbf{M}$ is a constant matrix.

2. Given two irreps $\mathbf{D}^{(1)}(R)$ and $\mathbf{D}^{(2)}(R)$ of dimensions l_1 and l_2 respectively, and a rectangular matrix $\mathbf{M}$ such that

$$\mathbf{M}\mathbf{D}^{(1)}(R) = \mathbf{D}^{(2)}(R)\mathbf{M} \quad \text{for all } R$$

then if $l_1 \neq l_2$, $\mathbf{M} \equiv 0$. On the other hand, if $l_1 = l_2$, either $\mathbf{M} \equiv 0$ or the two irreps are equivalent (i.e. they are the same irrep differing only by a similarity transformation).

Appendix E

The Symmetry–Coordinate Transformation tables for the 32 point groups and two infinite groups

E–1. *The Symmetry–Coordinate Transformation tables for the 32 point groups and two infinite groups. Included are the components of a polar vector* (x, y, z)*, an axial vector* (R_x, R_y, R_z)*, a second-rank symmetric tensor* ($\alpha_1, \ldots, \alpha_6$) *and of triple products of polar vector components* ($\beta_1, \ldots, \beta_{10}$)*. (See Eqs. (4–4) and Eq. (9–65) or page 67 for the definitions of the* α_i *and* β_i *components.)*

1. The nonaxial groups

C_1		
A	x; y; z; R_x; R_y; R_z	all α_i; all β_i

C_S			
A′	x; y; R_z	α_1; α_2; α_3; α_6	β_1; β_2; β_4; β_7; β_8; β_9
A″	z; R_x; R_y	α_4; α_5	β_3; β_5; β_6; β_{10}

C_i		
A_g	R_x; R_y; R_z	all α_i
A_u	x;y;z	all β_i

2. *The* C_n *groups*

C_2			
A	z; R_z	α_1; α_2; α_3; α_6	β_3; β_5; β_6; β_{10}
B	x; y; R_x; R_y	α_4; α_5	β_1; β_2; β_4; β_7; β_8; β_9

C_3			
A	z; R_z	$\alpha_1 + \alpha_2$; α_3	β_3; $\beta_5 + \beta_6$; $\beta_1 - \sqrt{3}\beta_7$; $\beta_2 - \sqrt{3}\beta_4$
$\tilde{E}$	$x \mp iy$; $R_x \mp iR_y$	$\alpha_4 \pm i\alpha_5$; $\alpha_1 - \alpha_2 \pm i\sqrt{2}\alpha_6$	$\beta_1 + \beta_7/\sqrt{3} \mp i(\beta_2 + \beta_4/\sqrt{3})$; $\beta_8 \mp i\beta_9$; $\beta_5 - \beta_6 \pm i\sqrt{2}\beta_{10}$

C_4			
A	z; R_z	$\alpha_1 + \alpha_2$; α_3	β_3; $\beta_5 + \beta_6$
B	–	$\alpha_1 - \alpha_2$; α_6	$\beta_5 - \beta_6$; β_{10}
$\tilde{E}$	$x \mp iy$; $R_x \mp iR_y$	$\alpha_4 \pm i\alpha_5$	$\beta_8 \mp i\beta_9$; $\beta_4 \pm i\beta_7$; $\beta_1 \mp i\beta_2$

C_6			
A	z; R_z	$\alpha_1 + \alpha_2$; α_3	β_3; $\beta_5 + \beta_6$
B	–	–	$\beta_1 - \sqrt{3}\beta_7$; $\beta_2 - \sqrt{3}\beta_4$
$\tilde{E}_1$	$x \mp iy$; $R_x \mp iR_y$	$\alpha_4 \pm i\alpha_5$	$\beta_1 + \beta_7/\sqrt{3} \mp i(\beta_2 + \beta_4/\sqrt{3})$; $\beta_8 \mp i\beta_9$
$\tilde{E}_2$	–	$\alpha_1 - \alpha_2 \mp i\sqrt{2}\alpha_6$	$\beta_5 - \beta_6 \mp i\sqrt{2}\beta_{10}$

3. *The* C_{nh} *groups*

C_{2h}		
A_g	R_z	$\alpha_1; \alpha_2; \alpha_3; \alpha_6$
B_g	$R_x; R_y$	$\alpha_4; \alpha_5$
A_u	z	$\beta_3; \beta_5; \beta_6; \beta_{10}$
B_u	$x; y$	$\beta_1; \beta_2; \beta_4; \beta_7; \beta_8; \beta_9$

C_{3h}			
A'	R_z	$\alpha_1 + \alpha_2; \alpha_3$	$\beta_1 - \sqrt{3}\beta_7; \beta_2 - \sqrt{3}\beta_4$
$\tilde{E}'$	$x \mp iy$	$\alpha_1 - \alpha_2 \pm i\sqrt{2}\alpha_6$	$\beta_8 \mp i\beta_9; \beta_1 + \beta_7/\sqrt{3} \mp i(\beta_2 + \beta_4/\sqrt{3})$
A''	z	–	$\beta_5 + \beta_6; \beta_3$
$\tilde{E}''$	$R_x \mp iR_y$	$\alpha_4 \pm i\alpha_5$	$\beta_5 - \beta_6 \pm i\sqrt{2}\beta_{10}$

C_{4h}		
A_g	R_z	$\alpha_1 + \alpha_2; \alpha_3$
B_g	–	$\alpha_1 - \alpha_2; \alpha_6$
$\tilde{E}_g$	$R_x \mp iR_y$	$\alpha_4 \pm i\alpha_5$
A_u	z	$\beta_3; \beta_5 + \beta_6$
B_u	–	$\beta_5 - \beta_6; \beta_{10}$
$\tilde{E}_u$	$x \mp iy$	$\beta_8 \mp i\beta_9; \beta_1 \mp i\beta_2; \beta_4 \pm i\beta_7$

C_{6h}		
A_g	R_z	$\alpha_1 + \alpha_2; \alpha_3$
B_g	–	–
$\tilde{E}_{1g}$	$R_x \mp iR_y$	$\alpha_4 \pm i\alpha_5$
$\tilde{E}_{2g}$	–	$\alpha_1 - \alpha_2 \mp i\sqrt{2}\alpha_6$
A_u	z	$\beta_3; \beta_5 + \beta_6$
B_u	–	$\beta_1 - \sqrt{3}\beta_7; \beta_2 - \sqrt{3}\beta_4$
$\tilde{E}_{1u}$	$x \mp iy$	$\beta_8 \mp i\beta_9; \beta_1 + \beta_7/\sqrt{3} \mp i(\beta_2 + \beta_4/\sqrt{3})$
$\tilde{E}_{2u}$	–	$\beta_5 - \beta_6 \mp i\sqrt{2}\beta_{10}$

4. The S_n groups

S_4			
A	R_z	$\alpha_1 + \alpha_2$; α_3	$\beta_5 - \beta_6$; β_{10}
B	z	$\alpha_1 - \alpha_2$; α_6	β_3; $\beta_5 + \beta_6$
$\tilde{E}$	$x \mp iy$; $R_x \pm iR_y$	$\alpha_4 \mp i\alpha_5$	$\beta_8 \mp i\beta_9$; $\beta_4 \pm i\beta_7$; $\beta_1 \mp i\beta_2$

S_6		
A_g	R_z	$\alpha_1 + \alpha_2$; α_3
$\tilde{E}_g$	$R_x \mp iR_y$	$\alpha_1 - \alpha_2 \pm i\sqrt{2}\alpha_6$; $\alpha_4 \pm i\alpha_5$
A_u	z	β_3; $\beta_5 + \beta_6$; $\beta_1 - \sqrt{3}\beta_7$; $\beta_2 - \sqrt{3}\beta_4$
$\tilde{E}_u$	$x \mp iy$	same as $\tilde{E}$ in group C_3

5. *The* C_{nv} *groups*

C_{2v}			
A_1	z	$\alpha_1; \alpha_2; \alpha_3$	$\beta_3; \beta_5; \beta_6$
A_2	R_z	α_6	β_{10}
B_1	$x; R_y$	α_5	$\beta_1; \beta_7; \beta_8$
B_2	$y; R_x$	α_4	$\beta_2; \beta_4; \beta_9$

C_{3v}			
A_1	z	$\alpha_1 + \alpha_2; \alpha_3$	$\beta_3; \beta_5 + \beta_6; \beta_2 - \sqrt{3}\beta_4$
A_2	R_z	–	$\beta_1 - \sqrt{3}\beta_7$
E	$(y, x); (R_x, -R_y)$	$(\alpha_4, \alpha_5); (\alpha_1 - \alpha_2, \alpha_6)$	$(\beta_2 + \beta_4/\sqrt{3}, \beta_1 + \beta_7/\sqrt{3});$
			$(\beta_9, \beta_8); (\beta_5 - \beta_6, \beta_{10})$

C_{4v}			
A_1	z	$\alpha_1 + \alpha_2; \alpha_3$	$\beta_3; \beta_5 + \beta_6$
A_2	R_z	–	–
B_1	–	$\alpha_1 - \alpha_2$	$\beta_5 - \beta_6$
B_2	–	α_6	β_{10}
E	$(y, x); (R_x - R_y)$	(α_4, α_5)	$(\beta_2, \beta_1); (\beta_4, \beta_7); (\beta_9, \beta_8)$

C_{6v}			
A_1	z	$\alpha_1 + \alpha_2; \alpha_3$	$\beta_3; \beta_5 + \beta_6$
A_2	R_z	–	–
B_1	–	–	$\beta_1 - \sqrt{3}\beta_7$
B_2	–	–	$\beta_2 - \sqrt{3}\beta_4$
E_1	$(y, x); (R_x, -R_y)$	(α_4, α_5)	$(\beta_9, \beta_8); (\beta_2 + \beta_4/\sqrt{3}, \beta_1 + \beta_7/\sqrt{3})$
E_2	–	$(\alpha_1 - \alpha_2, \alpha_6)$	$(\beta_5 - \beta_6, \beta_{10})$

6. The D_n *groups*

D_2			
A	–	α_1; α_2; α_3	β_{10}
B_1	z; R_z	α_6	β_3; β_5; β_6
B_2	y; R_y	α_5	β_2; β_4; β_9
B_3	x; R_x	α_4	β_1; β_7; β_8

D_3			
A_1	–	$\alpha_1 + \alpha_2$; α_3	$\beta_1 - \sqrt{3}\beta_7$
A_2	z; R_z	–	β_3; $\beta_5 + \beta_6$; $\beta_2 - \sqrt{3}\beta_4$
E	$(x, -y)$; $(R_x, -R_y)$	(α_4, α_5); $(\alpha_1 - \alpha_2, \alpha_6)$	$(\beta_1 + \beta_7/\sqrt{3}, -\beta_2 - \beta_4/\sqrt{3})$; $(\beta_8, -\beta_9)$; $(\beta_{10}, \beta_6 - \beta_5)$

D_4			
A_1	–	$\alpha_1 + \alpha_2$; α_3	–
A_2	z, R_z	–	β_3; $\beta_5 + \beta_6$
B_1	–	$\alpha_1 - \alpha_2$	β_{10}
B_2	–	α_6	$\beta_5 - \beta_6$
E	$(x, -y)$; $(R_x, -R_y)$	(α_4, α_5)	$(\beta_1, -\beta_2)$; $(\beta_8, -\beta_9)$; $(\beta_7, -\beta_4)$

D_6			
A_1	–	$\alpha_1 + \alpha_2$; α_3	–
A_2	z, R_z	–	β_3; $\beta_5 + \beta_6$
B_1	–	–	$\beta_1 - \sqrt{3}\beta_7$
B_2	–	–	$\beta_2 - \sqrt{3}\beta_4$
E_1	$(x, -y)$; $R_x, -R_y)$	(α_4, α_5)	$(\beta_1 + \beta_7/\sqrt{3}, -\beta_2 - \beta_4/\sqrt{3})$; $(\beta_8, -\beta_9)$
E_2	–	$(\alpha_1 - \alpha_2, \alpha_6)$	$(\beta_{10}, \beta_6 - \beta_5)$;

7. *The* D_{nh} *groups*

D_{2h}		
A_g	–	α_1; α_2; α_2
B_{1g}	R_z	α_6
B_{2g}	R_y	α_5
B_{3g}	R_x	α_4
A_u	–	β_{10}
B_{1u}	z	β_3; β_5; β_6
B_{2u}	y	β_2; β_4; β_9
B_{3u}	x	β_1; β_7; β_8

D_{4h}		
A_{1g}	–	$\alpha_1 + \alpha_2$; α_3
A_{2g}	R_z	–
B_{1g}	–	$\alpha_1 - \alpha_2$
B_{2g}	–	α_6
E_g	$(R_x, -R_y)$	(α_4, α_5)
A_{1u}	–	β's as in D_4 (all u irreps)
A_{2u}	z	
B_{1u}	–	
B_{2u}	–	
E_u	(x, y)	

D_{3h}			
A_1'	–	$\alpha_1 + \alpha_2$; α_3	$\beta_2 - \sqrt{3}\beta_4$
A_2'	R_z	–	$\beta_1 - \sqrt{3}\beta_7$
E'	(y, x)	$(\alpha_1 - \alpha_2, \alpha_6)$	$(\beta_2 + \beta_4/\sqrt{3}, \beta_1 + \beta_7/\sqrt{3})$; (β_9, β_8)
A_1''	–	–	–
A_2''	z	–	β_3; $\beta_5 + \beta_6$
E''	$(R_x, -R_y)$	(α_4, α_5)	$(\beta_5 - \beta_6, \beta_{10})$

D_{6h}		
A_{1g}	–	$\alpha_1 + \alpha_2$; α_3
A_{2g}	R_z	–
B_{1g}	–	–
B_{2g}	–	–
E_{1g}	$(R_x, -R_y)$	(α_4, α_5)
E_{2g}	–	$(\alpha_1 - \alpha_2, \alpha_6)$
A_{1u}	–	β's as in D_6 (all u irreps)
A_{2u}	z	
B_{1u}	–	
B_{2u}	–	
E_{1u}	(x, y)	
E_{2u}	–	

8. *The* D_{nd} *groups*

D_{2d}			
A_1	–	$\alpha_1 + \alpha_2$; α_3	β_{10}
A_2	R_z	–	$\beta_5 - \beta_6$
B_1	–	$\alpha_1 - \alpha_2$	–
B_2	z	α_6	β_3; $\beta_5 + \beta_6$
E	(x, y); $(R_x, -R_y)$	(α_4, α_5)	(β_1, β_2); (β_7, β_4); (β_8, β_9)

D_{3d}		
A_{1g}	–	$\alpha_1 + \alpha_2$; α_3
A_{2g}	R_z	–
E_g	$(R_x, -R_y)$	$(\alpha_1 - \alpha_2, \alpha_6)$; (α_4, α_5)
A_{1u}	–	β's as in D_3 (all u irreps)
A_{2u}	z	
E_u	(x, y)	

9. The cubic groups

T			
A	–	$\alpha_1 + \alpha_2 + \alpha_3$	β_{10}
$\tilde{\mathrm{E}}$	–	$2\alpha_1 - \alpha_2 - \alpha_3 \mp i\sqrt{3}(\alpha_2 - \alpha_3)$	–
T	(x, y, z); (R_x, R_y, R_z)	$(\alpha_4, \alpha_5, \alpha_6)$	$(\beta_1, \beta_2, \beta_3)$; $(\beta_7, \beta_9, \beta_5)$; $(\beta_8, \beta_4, \beta_6)$

T_h		
A_g	–	$\alpha_1 + \alpha_2 + \alpha_3$
$\tilde{E}_g$	–	$2\alpha_1 - \alpha_2 - \alpha_3 \mp i\sqrt{3}(\alpha_2 - \alpha_3)$
T_g	(R_x, R_y, R_z)	$(\alpha_4, \alpha_5, \alpha_6)$
A_u	–	β_{10}
$\tilde{E}_u$	–	–
T_u	(x, y, z)	$(\beta_1, \beta_2, \beta_3)$; $(\beta_7, \beta_9, \beta_5)$; $(\beta_8, \beta_4, \beta_6)$

T_d			
A_1	–	$\alpha_1 + \alpha_2 + \alpha_3$	β_{10}
A_2	–	–	–
E	–	$(2\alpha_1 - \alpha_2 - \alpha_3, \alpha_2 - \alpha_3)$	–
T_1	(R_x, R_y, R_z)	–	$(\beta_7 - \beta_8, \beta_9 - \beta_4, \beta_5 - \beta_6)$
T_2	(x, y, z)	$(\alpha_4, \alpha_5, \alpha_6)$	$(\beta_1, \beta_2, \beta_3)$; $(\beta_7 + \beta_8, \beta_9 + \beta_4, \beta_5 + \beta_6)$

O			
A_1	–	$\alpha_1 + \alpha_2 + \alpha_3$	–
A_2	–	–	β_{10}
E	–	$(2\alpha_1 - \alpha_2 - \alpha_3, \alpha_2 - \alpha_3)$	–
T_1	(x, y, z); (R_x, R_y, R_z)	–	$(\beta_1, \beta_2, \beta_3)$; $(\beta_7 + \beta_8, \beta_9 + \beta_4, \beta_5 + \beta_6)$
T_2	–	$(\alpha_4, \alpha_5, \alpha_6)$	$(\beta_7 - \beta_8, \beta_9 - \beta_4, \beta_5 - \beta_6)$

O_h		
A_{1g}	–	$\alpha_1 + \alpha_2 + \alpha_3$
A_{2g}	–	–
E_g	–	$(2\alpha_1 - \alpha_2 - \alpha_3, \alpha_2 - \alpha_3)$
T_{1g}	(R_x, R_y, R_z)	–
T_{2g}	–	$(\alpha_4, \alpha_5, \alpha_6)$
A_{1u}	–	β's as in O (all u irreps)
A_{2u}	–	
E_u	–	
T_{1u}	(x, y, z)	
T_{2u}	–	

10. The infinite groups (β's not included)

$C_{\infty v}$		
A_1	z	$\alpha_1 + \alpha_2; \alpha_3$
A_2	R_z	–
E_1	$(y, x); (R_x, -R_y)$	(α_4, α_5)
E_2	–	$(\alpha_1 - \alpha_2, \alpha_6)$
⋮	⋮	⋮

R(3)		
A	–	$\alpha_1 + \alpha_2 + \alpha_3$
T	$(x, y, z); (R_x, R_y, R_z)$	–
H	–	$(2\alpha_1 - \alpha_2 - \alpha_3, \alpha_2 - \alpha_3, \alpha_4, \alpha_5, \alpha_6)$
⋮		

Appendix F

Proof of the Fundamental Theorem

We begin with the expression for the relation between two physical quantities **X** and **Y** expressed as hypervectors in symmetry coordinates:

$$\bar{\mathbf{Y}} = \bar{\mathbf{K}}\bar{\mathbf{X}} \tag{F–1}$$

where $\bar{\mathbf{K}}$ is the corresponding matter tensor in symmetrized form. It is further assumed that all symmetry coordinates of $\bar{\mathbf{X}}$ and $\bar{\mathbf{Y}}$ that belong to a given two- or three-dimensional irrep are similarly oriented.

Consider, now, the effect of carrying out a symmetry operation R. The quantities $\bar{\mathbf{X}}$ and $\bar{\mathbf{Y}}$ transform as follows:

$$R\bar{\mathbf{X}} = \mathbf{D}_X(R)\bar{\mathbf{X}}; \quad R\bar{\mathbf{Y}} = \mathbf{D}_Y(R)\bar{\mathbf{Y}} \tag{F–2}$$

where $\mathbf{D}_X(R)$ and $\mathbf{D}_Y(R)$ are the reps of **X** and **Y** in block form. Equation (1) then becomes

$$\mathbf{D}_Y(R)\bar{\mathbf{Y}} = \bar{\mathbf{K}}'\mathbf{D}_X(R)\bar{\mathbf{X}} \tag{F–3}$$

where $\bar{\mathbf{K}}'$ is the new form of $\bar{\mathbf{K}}$. However, by Neumann's principle $\bar{\mathbf{K}}' = \bar{\mathbf{K}}$. Further, recognizing that $\mathbf{D}_Y(R)$ is unitary, we obtain

$$\bar{\mathbf{K}}' = \mathbf{D}_Y(R)^{\dagger}\bar{\mathbf{K}}\mathbf{D}_X(R)\bar{\mathbf{X}} = \bar{\mathbf{K}} \tag{F–4}$$

so that

$$\mathbf{D}_Y(R)\bar{\mathbf{K}} = \bar{\mathbf{K}}\mathbf{D}_X(R) \tag{F–5}$$

Recall that $\mathbf{D}_Y(R)$ and $\mathbf{D}_X(R)$ are in block form. It is possible, then, to arrange $\bar{\mathbf{K}}$ into corresponding blocks. To illustrate, consider the case in which $\Gamma_Y = 2\alpha + \beta$ and $\Gamma_X = \alpha$, where α and β are irreps. Equation (5) in block form then becomes:

$$\underbrace{\begin{bmatrix} \alpha & 0 & 0 \\ 0 & \alpha & 0 \\ 0 & 0 & \beta \end{bmatrix}}_{\mathbf{D}_Y(R)} \underbrace{\begin{bmatrix} (\alpha\alpha)_1 \\ (\alpha\alpha)_2 \\ \alpha\beta \end{bmatrix}}_{\bar{\mathbf{K}}} = \underbrace{\begin{bmatrix} (\alpha\alpha)_1 \\ (\alpha\alpha)_2 \\ \alpha\beta \end{bmatrix}}_{\bar{\mathbf{K}}} \underbrace{\begin{bmatrix} \alpha \end{bmatrix}}_{\mathbf{D}_X(R)}$$

Here the blocks α of $\mathbf{D}_Y(R)$ and $\mathbf{D}_X(R)$ are all identical, while two of the three blocks of $\bar{\mathbf{K}}$ are those connecting α and α and the remaining one, α and β. The blocks of $\bar{\mathbf{K}}$ have appropriate dimensions commensurate with those of α and β.

To complete the proof, we refer to the two lemmas used to prove the Great Orthogonality Theorem (Appendix D). Schur's lemma tells us that all blocks in $\bar{\mathbf{K}}$ of the type $\alpha\alpha$ are constant submatrices. The second lemma tells us that all blocks of type $\alpha\beta$ (i.e. relating different irreps) are zero submatrices. The possibility of equivalence mentioned in the second lemma is ruled out by the use of similarly oriented symmetry coordinates, which guarantees that all blocks of a given irrep in both $\mathbf{D}_Y(R)$ and $\mathbf{D}_X(R)$ must have *identical* (not just equivalent) forms. From this result, the two statements of the Fundamental Theorem immediately follow.

Appendix G

Theorems concerning magnetic groups

Theorem 1. All magnetic groups of type III can be written in the form $\mathcal{M} = \mathcal{H} + \theta(\mathcal{G} - \mathcal{H})$, or $\mathcal{M} = \{A_i\} + \{\underline{B}_i\}$ in which $\mathcal{H}$ is a subgroup of $\mathcal{G}$ of order g/2.

Proof. Given a group with a set of elements $\{A_i\}$ and at least one operation $\underline{B}_1 = \theta B_1$ in the set $\{\underline{B}_j\}$ and with θ not an element of the group $\mathcal{G}$. First we show that

$$A_i A_j = A_k \tag{G–1}$$

that is, that the product of two members of the $\{A_i\}$ set give another member of that set. To show this, we assume the contrary, namely, $A_1 A_2 = \underline{B}_1 = \theta B_1$ where B_1 is an ordinary symmetry operation in the original group $\mathcal{G}$. Then

$$\theta = A_1 A_2 B_1^{-1}$$

must be a member of group $\mathcal{G}$, which is not the case. Thus, Eq. (1) is valid. Next, we show that

$$\underline{B}_i \underline{B}_j = A_k \tag{G–2}$$

that is, that the product of any two members of the $\{\underline{B}_j\}$ set is a member of $\{A_i\}$. This follows from the fact that

$$\underline{B}_i \underline{B}_j = \theta B_i \theta B_j = \theta^2 B_i B_j = B_i B_j$$

since θ commutes with all ordinary symmetry operations and $\theta^2 = E$. The resulting product $B_i B_j$ must be a non-complementary member of $\mathcal{G}$, therefore a member of $\{A_i\}$. Finally, in a similar way, we show that

$$A_i \underline{B}_j = \underline{B}_k \tag{G–3}$$

that is, that the product of a member of $\{A_i\}$ and a member of $\{\underline{B}_j\}$ is another member of $\{\underline{B}_j\}$.

From Eq. (1), it is clear that the set $\{A_i\}$ forms a group, since it has closure, it includes the identity operation E and includes the inverse of each operation: $A_i^{-1} A_i = E$. Thus we have $\mathcal{M}$ made up of $\{A_i\}$ as a subgroup of $\mathcal{G}$ in addition to a set $\{B_j\}$. Say there are g' members of $\{A_i\}$.

We now show that $\{B_j\}$ contains at least as many members as $\{A_i\}$. Since $\{B_j\}$ has at least one element $\underline{B}_1$, consider the set $\{\underline{B}_1 A_i\}$. From Eq. (3), this contains g' different elements all of the type $\underline{B}_j$. On the other hand, the number of $\{\underline{B}_j\}$

cannot be greater than g'; for if it were, Eq. (2) would require that

$$\underline{B}_1\underline{B}_j = \underline{B}_1\underline{B}_k$$

for some j and k, leading to $B_j = B_k$, which is a contradiction (since B_j and B_k are both members of the original group $\mathcal{G}$). It is concluded that the subgroup $\{A_i\}$ has exactly $g/2$ elements, as does the set $\{\underline{B}_j\}$.

From the definition of type-III magnetic groups, it is now clear that such a group $\mathcal{M}$ can be formed from a group $\mathcal{G}$ (of order g) by selecting any subgroup $\mathcal{H}$ of order $g/2$ as the set $\{A_i\}$ and placing the remaining $g/2$ operations into the set $\{\underline{B}_j\}$. Thus, every subgroup $\mathcal{H}$ of order $g/2$ can generate a magnetic group, which we designate as $\mathcal{G}$: $\mathcal{H}$. □

Theorem 2. Given a classical group $\mathcal{G}$ (of order g) composed of $\{A_i\} + \{B_j\}$ where $\{A_i\}$ constitute a subgroup of order $g/2$, consider the irreps of $\mathcal{G}$. For every irrep γ, there exists an irrep γ^c, called the *complementary irrep of* γ, such that

$$\left.\begin{aligned} \chi^c(A_i) &= \chi(A_i) \\ \chi^c(B_j) &= -\chi(B_j) \end{aligned}\right\} \qquad \text{(G–4)}$$

where $\chi(A_i)$ and $\chi(B_j)$ are the characters of γ for operations A_i and B_j, respectively, and $\chi^c(A_i)$ and $\chi^c(B_j)$ are the corresponding characters for the complementary irrep.

Proof. Let $\mathbf{D}_\gamma(R)$ be the irrep matrices for irrep γ of group $\mathcal{G}$. These matrices obey the same multiplication table as that of the group operations (as given by Eqs. (1)–(3), which are equally valid without bars below the B_j). Now define γ^c by a set of matrices as follows:

$$\left.\begin{aligned} \mathbf{D}^c_\gamma(A_i) &= \mathbf{D}_\gamma(A_i) \\ \mathbf{D}^c_\gamma(B_j) &= -\mathbf{D}_\gamma(B_j) \end{aligned}\right\} \qquad \text{(G–5)}$$

These complementary matrices also constitute a rep of $\mathcal{G}$, since they obey the same multiplication table of the operations of $\mathcal{G}$ as do the original matrices $\mathbf{D}_\gamma(R)$. Now, if $\mathbf{D}_\gamma(R)$ is a one-dimensional irrep, so is $\mathbf{D}^c_\gamma(R)$. Also, if $\mathbf{D}_\gamma(R)$ is two- or three-dimensional, $\mathbf{D}^c_\gamma(R)$ is also an irrep. The test for an irrep is that

$$\sum_R \chi^*(R)\chi(R) = g \qquad \text{(G–6)}$$

as is evident from the orthogonality theorem, Eq. (3–13). Clearly, Eq. (6) is equally true for γ^c as it is for γ. □

Theorem 3. Every magnetic group of $\mathcal{G}$ can be obtained from a one-dimensional irrep of $\mathcal{G}$ (other than the totally symmetric irrep and any $\tilde{\text{E}}$ irreps) by letting $\{A_i\}$ and $\{B_j\}$ constitute those group operations for which the characters are +1 and −1, respectively.

Proof. Such an assignment of $\{A_i\}$ and $\{B_j\}$ is consistent with Theorem 1, namely, half of the characters must be +1 and half = −1 in order to meet the requirement of orthogonality to the totally symmetric irrep (see Eq. (3–13), in which $\chi^{(\alpha)}(R)^*$ may all be set = 1). Furthermore, the set $\{A_i\}$ forms a subgroup, in that the A_i obey Eq. (1) representing closure, and it also includes the identity operation, and the inverse of each A_i.

Conversely, given a magnetic group $\mathcal{M}$ of $\mathcal{G}$ made up of $\{A_i\}$ and $\{\underline{B}_j\}$. Assign $+1$ to all operations A_i and -1 to all B_j, as is done in Table 5–1. This set of numbers constitutes a rep (in view of Eqs. (1)–(3) with B_j replacing $\underline{B}_j$). At the same time, it is an irrep, since it is one-dimensional. Thus, this theorem of matching magnetic groups to irreps is valid for all magnetic groups and all irreps that have ± 1 characters. □

References

Chapter 1

Bhagavantam, S. (1966) *Crystal Symmetry and Physical Properties*, Academic, London.

Landau, L. D. and Lifschitz, E.M. (1960), *Electrodynamics of Continuous Media*, Addison-Wesley, Reading, Chapter 11.

Mason, W.P. (1966), *Crystal Physics of Interaction Processes*, Academic, New York.

Narasimhamurty, T. S. (1981), *Photoelastic and Electro-optic Properties of Crystals*, Plenum, New York.

Nye, J.F. (1957), *Physical Properties of Crystals*, Oxford, London.

Truell, R., Elbaum, C. and Chick, B.B. (1969), *Ultrasonic Methods in Solid State Physics*, Academic, New York.

Wallace, D.C. (1970), Thermoelastic theory of stressed crystals and higher order elastic constants, *Solid-State Phys.* **25**, 302.

Chapter 2

de Groot, S. R. and Mazur, P. (1962), *Non-equilibrium Thermodynamics*, North Holland, Amsterdam.

Denbigh, K. G. (1951), *Thermodynamics of the Steady State*, Methuen, London.

Haase, P. (1969), *Thermodynamics of Irreversible Processes*, Dover, New York.

Mason, W. P. (1966), *Crystal Physics of Interaction Processes*, Academic, New York, Chapters 9–12.

Prigogene, I. (1961), *Introduction to Thermodynamics of Irreversible Processes*, 2nd edn, Interscience, New York.

Chapter 3

Bishop, D. M. (1973), *Group Theory and Chemistry*, Clarendon Press, Oxford.

Burns, G. (1977), *Introduction to Group Theory with Applications*, Academic Press, New York.

Janot, C. (1992), *Quasicrystals*, Clarendon Press, Oxford.

Tinkham, M. (1964), *Group Theory and Quantum Mechanics*, McGraw-Hill, New York.
Wilson, E. B, Decius, J. C. and Cross, P. C. (1955), *Molecular Vibrations*, McGraw-Hill, New York.

Chapter 4

Fumi, F. G. (1952a), Physical properties of crystals: The direct-inspection method, *Acta Cryst.* **5**, 44 and 691.
Fumi, F. G. (1952b), Matter tensors in symmetrical systems, *Il Nuovo Cimento* **9**, 739.
Juretschke, H. J. (1974), *Crystal Physics*, W. A. Benjamin, Reading, Chapter 4.
Nowick, A. S. and Heller, W. R. (1965), Dielectric and anelastic relaxation of crystals containing point defects, *Advan. Phys.* **14**, 101.
Nye, J. F. (1957), *Physical Properties of Crystals*, Oxford University Press, London.
Weyl, H. (1950), *Theory of Groups and Quantum Mechanics*, Dover, New York, Chapter III.
Wilson, E. G., Decius, J. C. and Cross P. C. (1955), *Molecular Vibrations*, McGraw-Hill, New York.

Chapter 5

Birss, R. R. (1964), *Symmetry and Magnetism*, North-Holland, Amsterdam, Chapters 1–3.
Bhagavantum, S. (1966), *Crystal Symmetry and Physical Properties*, Academic Press, London, Chapter 5.
Cracknell, A. P. (1975), *Group Theory in Solid State Physics*, Taylor and Francis, London, Chapters 3–4.
Shubnikov, A. V. and Belov, N. V. (1964), *Colored Symmetry*, Pergamon Press, Oxford.

Chapter 6

Hagenmuller, P. and van Gool, W. (eds.) (1978), *Solid Electrolytes*, Academic, New York, Chapters 21–3.
International Tables of X-ray Crystallography (1952), Kynoch Press, Birmingham, Volume I.
Johnson, O. W. (1964), One-dimensonal diffusion of lithium in rutile, *Phys. Rev.* **A136**, 284.
Jona, F. and Shirane, G. (1962), *Ferroelectric Crystals*, Pergamon, New York.
Lines, M. E. and Glass, A. M. (1977), *Principles and Applications of Ferroelectrics and Related Materials*, Clarendon Press, Oxford.
Mason, W. P. (1966), *Crystal Physics of Interaction Processes*, Academic Press, New York, Chapters 7–9, 11.
Rothman, S. J., Routbort, J. L., Welp, U. and Baker, J. E. (1991), Anisotropy of oxygen tracer diffusion in single-crystal $YBa_2Cu_3O_{7-x}$, *Phys. Rev.* **B44**, 2326.

Shuvalov, L. A. (ed.) (1988), *Modern Crystallography IV*, Springer-Verlag, Berlin, Chapters 3, 5–7.
Tozer, S. W., Kleinsasser, A. W., Penney, T., Kaiser, D. and Holtzberg, F. (1987), Measurement of anisotropic resistivity and Hall constant for single-crystal $YBa_2Cu_3O_{7-x}$, *Phys. Rev. Lett.* **59**, 1768.

Chapter 7

Buchanan, R. C. (ed.) (1991), *Ceramic Materials for Electronics*, Marcel Dekker, New York, Chapter 3.
Cady, W. G. (1964), *Piezoelectricity*, Dover, New York, Volumes 1 and 2.
Günter, P. (ed.) (1987), *Electro-optic and Photorefractive Materials*, Springer, Berlin, Part I.
Mason, W. P. (1958), *Physical Acoustics and the Properties of Solids*, Van Nostrand, New York, p. 54.
Mason, W. P. (1966), *Crystal Physics of Interaction Processes*, Academic Press, New York, Chapters 6 and 7.
Narasimhamurty, T. S. (1981), *Photoelastic and Electro-optic Properties of Crystals*, Plenum, New York, Chapters 7–8.
Shuvalov, L. A. (1988). *Modern Crystallography IV*, Springer-Verlag, Berlin, Section 3.3.
Wiesenfeld, J. M. (1990), Electro-optic sampling of high-speed devices and integrated circuits, *IBM J. Res. Dev.* **34**, 141.

Chapter 8

Bhagavantam, S. (1966), *Crystal Symmetry and Physical Properties*, Academic Press, London, Chapter 15.
Birss, R. R. (1964), *Symmetry and Magnetism*, North-Holland, Amsterdam, Chapter 4.
Mason, W. P. (1966), *Crystal Physics of Interaction Processes*, Academic, New York, Chapter 6.
Shuvalov, L. A. (ed.) (1988), *Modern Crystallography IV*, Springer-Verlag, Berlin, Chapter 4.

Chapter 9

Hearmon, R. F. S. (1946), The elastic constants of anisotropic materials, *Rev. Mod. Phys.* **18**, 409; (1956), *Phil. Mag. Suppl.* **5**, 323.
Jan, J. P. (1957), Galvanomagentic and themomagnetic effects in metals, *Solid-State Phys.* **5**, 3.
Keyes, R. W. (1960), Piezoresistance of semiconductors, *Solid-State Phys.* **11**, 149
Mason, W. P. (1964), Semiconductor transducers – general considerations, in: *Physical Acoustics* IB (ed. W. P. Mason), Academic Press, New York, Chapter 10.
Narasimhamurty, T. S. (1981), *Photoelastic and Electro-optic Properties of Crystals*, Plenum, New York, Chapter 5.
Nowick, A. S. and Berry, B. S. (1972), *Anelastic Relaxation in Crystallic Solids*, Academic Press, New York, Chapters 6 and 20.

Thurston, R. N. (1964), Use of semiconductor tranducers in measuring strains, accelerations and displacements, in: *Physical Acoustics* IB (ed. W. P. Mason), Academic Press, New York, Chapter 11.
Truell, R., Elbaum, C. and Chick, B.B. (1969), *Ultrasonic Methods in Solid State Physics*, Academic, New York, Chapters 1 and 2.

For numerical values of elastic constants as well as piezoelectric, electro-optic, and peizo-optic constants, see: Landolt–Börnstein Tables, Volume III/11, Springer-Verlag, Berlin (1979).

Chapter 10

Birss, R. R. (1964), *Symmetry and Magnetism*, Academic Press, London, Chapter 5.
Breazeale, M. A. and Philip, J. (1984), 'Determination of third-order elastic constants from ultrasonic harmonic generation measurements', in *Physical Acoustics*, Vol. 17, (eds. W. P. Mason and R. N. Thurston), Academic Press, Orlando, Chapter 1.
Brugger, K. (1965), 'Determination of third-order elastic coefficients in crystals', *J. Appl. Phys.* **36**, 759 and 768.
Burns, G. (1985) *Solid State Physics*, Academic Press, Orlando, Chapter 12.
Fieschi, R. and Fumi, F. G. (1953), 'Higher order matter tensors in symmetric systems', *Nuovo Cimento* **10**, 865.
Hearmon, R. F. S. (1953), 'Third-order elastic coefficients', *Acta Cryst.* **6**, 331.
Juretschke, H. J. (1974), *Crystal Physics*, W. A. Benjamin, Reading, Chapter 13.
Keyes, R. W. (1960), Piezoresistance of semiconductors, *Solid-State Phys.* **11**, 149
Thurston, R.N. (1964), 'Wave propagation in fluids and normal solids', in *Physical Acoustics* IA (ed. W. P. Mason), Academic Press, New York.
Wallace, D.C. (1970), Thermoelastic theory of stressed crystals and higher order elastic constants, *Solid-State Phys.* **25**, 302.

Appendix A

Hall, G. G. (1963), *Matrices and Tensors*, Pergamon, Oxford.

Appendices B and C

Bhagavantum, S. (1966), *Crystal Symmetry and Physical Properties*, Academic Press, London, Chapters 9–12.
Wallace, D.C. (1970), Themoelectric theory of stressed crystals and higher order elastic constants, *Solid-State Phys.* **25**, 302.

Index